Revu Alexander

Urokinase receptor (uPAR) and integrin interactions in Angiogenesis

Revu Alexander

Urokinase receptor (uPAR) and integrin interactions in Angiogenesis

Urokinase receptor and Integrin interplay on endothelial cells and its implication in VEGF-induced angiogenesis

Südwestdeutscher Verlag für Hochschulschriften

Impressum/Imprint (nur für Deutschland/only for Germany)
Bibliografische Information der Deutschen Nationalbibliothek: Die Deutsche Nationalbibliothek verzeichnet diese Publikation in der Deutschen Nationalbibliografie; detaillierte bibliografische Daten sind im Internet über http://dnb.d-nb.de abrufbar.
Alle in diesem Buch genannten Marken und Produktnamen unterliegen warenzeichen-, marken- oder patentrechtlichem Schutz bzw. sind Warenzeichen oder eingetragene Warenzeichen der jeweiligen Inhaber. Die Wiedergabe von Marken, Produktnamen, Gebrauchsnamen, Handelsnamen, Warenbezeichnungen u.s.w. in diesem Werk berechtigt auch ohne besondere Kennzeichnung nicht zu der Annahme, dass solche Namen im Sinne der Warenzeichen- und Markenschutzgesetzgebung als frei zu betrachten wären und daher von jedermann benutzt werden dürften.

Coverbild: www.ingimage.com

Verlag: Südwestdeutscher Verlag für Hochschulschriften GmbH & Co. KG
Heinrich-Böcking-Str. 6-8, 66121 Saarbrücken, Deutschland
Telefon +49 681 37 20 271-1, Telefax +49 681 37 20 271-0
Email: info@svh-verlag.de

Approved by: Wien, MUV, Diss., 2012

Herstellung in Deutschland (siehe letzte Seite)
ISBN: 978-3-8381-3340-9

Imprint (only for USA, GB)
Bibliographic information published by the Deutsche Nationalbibliothek: The Deutsche Nationalbibliothek lists this publication in the Deutsche Nationalbibliografie; detailed bibliographic data are available in the Internet at http://dnb.d-nb.de.
Any brand names and product names mentioned in this book are subject to trademark, brand or patent protection and are trademarks or registered trademarks of their respective holders. The use of brand names, product names, common names, trade names, product descriptions etc. even without a particular marking in this works is in no way to be construed to mean that such names may be regarded as unrestricted in respect of trademark and brand protection legislation and could thus be used by anyone.

Cover image: www.ingimage.com

Publisher: Südwestdeutscher Verlag für Hochschulschriften GmbH & Co. KG
Heinrich-Böcking-Str. 6-8, 66121 Saarbrücken, Germany
Phone +49 681 37 20 271-1, Fax +49 681 37 20 271-0
Email: info@svh-verlag.de

Printed in the U.S.A.
Printed in the U.K. by (see last page)
ISBN: 978-3-8381-3340-9

Copyright © 2012 by the author and Südwestdeutscher Verlag für Hochschulschriften GmbH & Co. KG and licensors
All rights reserved. Saarbrücken 2012

ACKNOWLEDGEMENTS

As I now stand at the precipice of this great odyssey, which is the PhD, I do agree. Science is not for the faint-hearted. One day you are at the top of the world brimming with exciting results and ideas, and the next day everything can come crashing down. But the entire experience was not only about learning scientific techniques, rational questioning and thinking out of the box. It was also about learning to believe in myself, to be strong and confident. And for that I am humbly grateful to many.

To late Prof.Bernd.B.Binder, who gave me the opportunity to pursue my interest in science. For giving "carte blanche" to try out ideas, and encouraging me to look at the bright side no matter what the outcome was. I hope I can emulate your spirit and passion for life and science.

To Prof.Michael Freissmuth, for the selflessness and generosity shown at difficult times. For his relentless enthusiasm to encourage students to think, ask and prode science. Hopefully I have imbibed some of his lessons.

To Prof. Johannes.M.Breuss, for the discussions scientific and otherwise; and especially the willingness to help out with any difficulties.

The Medical University of Vienna and Austrian Science Fund (FWF) who provided the financial support to my graduate studies through the CCHD (Cell Communication in Health and Disease Phd Program).

To Prof. Stefan Böhm as well as to the faculty of the CCHD program, who had excellently co-ordinated the CCHD program with journal clubs and seminars.

To Prof. Gerhard Schütz (Johannes Kepler University, Linz), who gave me the opportunity to do a six month external semester in his lab. Special mention to Stefan Sunzenauer, who patiently helped me with the scary microscopes.

To all the members of my lab, for creating a conducive environment, for the friendships and for all the help concerning reagents or instruments.

To all my friends in Vienna, for the good times which helped me to destress and have a life outside the lab.

Of course, my soul mates Srividya and Jaya. The two people whom I could constantly whine, cry, muse about life, reality and dreams. Inspite of being in different countries, you guys could always make me feel not alone. Ode to the internet.

Amulya. Teaching me to smile, be brave and to keep a positive attitude no matter what.

My two brothers: you guys will always be there for me.

My parents. I know that they are under the greatest peer pressure as to where their daugther's doings are taking her. Though being perplexed to my constant worrying about cells or experiments, they have always expressed a silent faith and support in my endeavours. I hope I won't let you down.

TABLE OF CONTENTS	PAGE
LIST OF ABBREVIATIONS	5
ZUSAMMENFASSUNG	7
ABSTRACT	9

1. INTRODUCTION

1.1 Vascular endothelial growth factor (VEGF)	12
1.2 The plasminogen system	15
1.3 Structure of uPAR and uPA	16
1.4 uPAR in angiogenesis, proteolysis and migration	18
1.5 Integrins	20
1.6 Integrins in angiogenesis	21
1.7 Interactions between uPAR and integrins	23
1.8 The LDL-receptor family	26
1.9 LDL-receptor mediated uPAR interactions	28
1.10 Rationale of the work	31

2. MATERIALS AND METHODS

2.1 Reagents and antibodies	32
2.2 Cell culture	33
2.3 Cytofluorimetric analysis	33
2.4 Surface biotinylation and Western blotting	34
2.5 siRNA treatment of endothelial cells	35
2.6 Immunocytochemistry	35
2.7 Micropatterning	35
2.8 Transwell migration assay	36
2.9 DIVAA in vivo angiogenesis assay	36

3. RESULTS

3.1 Internalization of uPAR and $\beta 1$ integrin upon VEGF stimulation.	38
3.2 uPAR is not cleaved during short term VEGF stimulation.	41
3.3 uPAR and $\beta 1$ integrin interact upon VEGF stimulation	42

3.4 uPAR and β1 integrin co-internalize in endocytic vesicles upon VEGF stimulation. **45**

3.5 VEGF induced PI3-kinase pathway modulates β1 integrin internalization. **48**

3.6 Domain 3 of uPAR is necessary for β1 integrin endocytosis. **50**

3.7 Integrin α5β1 associates with uPAR for VEGF induced endocytosis. **52**

3.8 UPAR is essential for VEGF induced internalization of α5β1 but not α3β1 integrin. **55**

3.9 The LDL-receptor is necessary for VEGF induced endocytosis of α5β1 integrin. **59**

3.10 Inhibition of uPAR-α5β1 interaction impairs VEGF induced migration of endothelial cells **60**

3.11 uPAR-integrin interaction is necessary for *in vivo* angiogenesis **62**

4. DISCUSSION **64**

5. BIBLIOGRAPHY **68**

LIST OF ABBREVIATIONS

ATF	amino terminal fragment
BSA	bovine serum albumin
ECM	extracellular matrix
FACS	fluorescence-activated cell sorting
GPI	glycosylphosphatidylinositol
HUVEC	human umbilical vein endothelial cell
LDL-R	low density lipoprotein receptor
LRP	LDL receptor related protein
MMP	matrix metalloproteinase
PAE	porcine aortic endothelial
PAI-1	plasminogen activator inhibitor-1
PBS	phosphate buffered saline
PDGF	platelet derived growth factor
PDGFR	platelet derived growth factor receptor
PI-PLC	phosphatidylinositol-specific phospholipase C
PLC-γ	phospholipase C-γ
PLGF	placental growth factor
PMA	phorbol-12-myristate-13-acetate
RAP	receptor-associated protein
TIR	total internal reflection
TIMP	tissue inhibitor of metalloproteinases
tPA	tissue plasminogen activator
uPA	urokinase plasminogen activator
uPAR	urokinase plasminogen activator receptor
VEGF	vascular endothelial growth factor
VEGFR	vascular endothelial growth factor receptor
VLDLR	very low-density lipoprotein receptor

ABBREVIATIONS FOR AMINO ACIDS.

Alanine	A	Glycine	G	Proline	P
Arginine	R	Histidine	H	Serine	S
Asparagine	N	Leucine	L	Threonine	T
Aspartic Acid	D	Isoleucine	I	Tryptophan	W
Cysteine	C	Lysine	K	Tyrosine	Y
Glutamine	Q	Methionine	M	Valine	V
Glutamic acid	E	Phenylalanine	F	Any aminoacid	X

ZUSAMMENFASSUNG

Vascular endothelial growth factor (VEGF) ist ein sehr potenter Wachstumsfaktor, der sowohl die physiologische als auch die pathologische Angiogenese steuert. VEGF initiiert Angiogenese, indem er die Migration von Endothelzellen stimuliert. Die stimulierten Endothelzellen dringen in das umgebende Gewebe ein. Dies erfordert, dass der proteolytische Abbau der extrazellulären Matrix durch das Plasminogen-System mit der gerichteten Zell-Migration durch Integrin-Matrix-Interaktionen koordiniert wird. Der Urokinase-Rezeptor (uPAR), der an der zelloberfläche exprimiert wird, ist eine zentrale Komponente des Plasminogen-Systems. Der Komplex aus aktiver Urokinase/Plasminogen-Aktivator (uPA) und dem Urokinase-Rezeptor (uPAR) beschleunigt die Umwandlung von Plasminogen zu Plasmin, damit wird eine proteolytische Kaskade aktiviert, die zum Abbau der extrazellulären Matrix führt. Frühere Versuche wiesen nach, dass eine Stimulation mit VEGF bei Endothelzellen nicht nur zur Aktivierung der pro-Urokinase (pro-uPA) und damit zu einem Abbau der extrazellulären Matrix führt sondern auch zu einer koordinierten Internalisierung des Urokinase-Rezeptors in einem Komplex mit einem Mitglied der LDLR-Familie führt. Dies wird durch die Umverteilung des Protease-Rezeptor uPAR an fokalen Adhäsionen ausgelöst. Die Bedeutung der Internalisierung ist offensichtlich: Wenn die Internalisierung von uPAR verhindert wird, wird die endotheliale Zellmigration unterdrückt und damit *in vivo* die Angiogenese gehemmt. Da uPAR keine direkten Liganden in der nativen Basalmembran hat, muss seine regulierte Internalisierung durch Assoziation mit einem anderen Zelloberflächenmolkül zustande kommen. Die Arbeitshypothese der vorliegenden Dissertation ging dacon aus, dass uPAR mit einem Integrin interagiert, weil Integrin-Matrix-Interaktionen eine wichtige Rolle bei der Zellwanderung spielen. Diese Hypothese wurde in verschiedenenAnsätzen geprüft: (i) Mittels Durchflusszytometrie wurde die Internalisierung von uPAR und β1-Integrinen in VEGF-stimulierten Endothelzellen quantifiziert. (ii) Mit konfokaler Mikroskopie wurde die VEGF-induzierte Cointernalisierung von uPAR und α5β1-Integrin in Endosomen visualisiert. (iii) Diese Internalisierung war abhängig von uPAR und von Rezeptoren aus der Low Density Lipoprotein (LDL)-Rezeptor-Familie. Die VEGF-induzierte Integrin-Umverteilung war in Abwesenheit von uPAR herabgesetzt; sie wurde auch durch inhibitorische Peptide, die die uPAR / Integrin-Interaktion blockieren, gehemmt. Wenn uPAR-defiziente Endothelzellen bzw. Endothelzellen, die mit inhibitorischen Peptiden inkubiert waren, mit VEGF stimuliert wurden, war die die Migration der Endothelzellen herabgesetzt. (iv) Die VEGF-induzierte Interaktion zwischen uPAR und α5β1-Integrin wurde an lebenden Endothelzellen mit einem neuartigen Ansatz der Mikrostrukturierung visualisiert: Mit immobilisierten gegen α5β1-Integrin gerichteten Antikörpern wurden die Integrine auf der endothelialen Zelloberfläch in ein Muster gezwungen. Nach Inkubation der Endothelzellen mit VEGF verteile sich uPAR in dasselbe Muster. Diese Beobachtung beweist, dass die beiden Proteine in lebenden Endothelzellen in Abhängigkeit von VEGF miteinander interagieren. Die Ergebnisse der Experimente liefern daher eine Bestätigung für die Arbeitshypothese. Darüber hinaus legen die Beobachtungen nahe, dass uPAR ein wesentlicher Bestandteil des Netzwerks an Signalproteinen, über die VEGF die Endothelzellmigration kontrolliert: uPAR ist ein Flaschenhals, durch den die VEGF-induzierte Signale geschleust werden müssen, um sowohl die proteolytische Aktivität an der invadierenden Membran zu fokussieren als um die Umverteilung und Rezirkulation der Integrine zu steuern.

ABSTRACT

Vascular endothelial growth factor (VEGF) is a very potent growth factor that drives both, physiological and pathological angiogenesis. VEGF initiates angiogenesis by activating endothelial cells to migrate and invade surrounding tissues. This requires the coordinated proteolytic degradation of the extracellular matrix by the plasminogen system and regulation of cell-migration provided by integrin-matrix interaction. The cell surface-bound urokinase receptor (uPAR) is a central component of the plasminogen system. The active urokinase:urokinase receptor complex accelerates the conversion of plasminogen to plasmin, initiating a proteolytic cascade leading to extracellular matrix degradation. Previous experiments showed that VEGF-stimulation of endothelial cells induced pro urokinase (pro-uPA) activation; this was accompanied not only by extracellular matrix degradation but also by coordinated internalization of the urokinase receptor complexed to a member of the LDLR family. Internalization is preceded by redistribution of the protease receptor uPAR to focal adhesions. Preventing the internalization of uPAR blocks endothelial cell migration and *in vivo* angiogenesis. uPAR does not have any primary ligand in the native basement membrane. Hence, internalization is likely to be contingent on the formation of uPAR with another partner. Because of the central role of integrins in matrix interactions, the working hypothesis underlying this thesis posits that uPAR forms a complex and is internalized with integrins. This conjecture was verified by using several approaches: (i) flow cytometry-based internalization experiments allowed for quantifying the VEGF-induced internalization of uPAR in endothelial cells. (ii) Confocal microscopy visualized the co-internalization of uPAR and $\alpha5\beta1$ integrins upon angiogenic stimulation of endothelial cells with VEGF. (iii) This retrieval was contingent on uPAR and on receptors of the low density lipoprotein (LDL) receptor family. VEGF-induced integrin redistribution was inhibited by elimination of uPAR from the endothelial cell surface or by inhibitory peptides that block the uPAR/integrin interaction. The absence of uPAR or the presence of the inhibitory peptides also imparied the migratory response of endothelial cells to VEGF. (iv) VEGF-promoted complex formation between uPAR and $\alpha5\beta1$ integrin was also visualized on live endothelial cells by using a micropatterning approach: immobilized $\alpha5\beta1$ integrin antibody imposed a pattern on the surface-expressed $\alpha5\beta1$ integrins. In response to VEGF uPAR was driven into the complex and thus adopted the pattern as dictated by the $\alpha5\beta1$ integrin antibody. This observation provided direct and incontrovertible evidence that the two proteins interacted in live VEGF-stimulated endothelial cells. Taken together the observations confirmed the working hypothesis. The findings indicate that uPAR is an essential component of the network through which VEGF controls endothelial cell migration: uPAR is a bottleneck through which the

VEGF-induced signal must be funnelled for both, focused proteolytic activity at the leading edge and for redistribution and recycling of integrins.

1. INTRODUCTION

Every blood vessel in our body, starting from the veins, arteries to the very small capillaries is lined by an endothelium. Endothelial cells do not only regulate the exchange of water, electrolytes and nutrients between blood and extracellular space and confine blood cells like the platelets or WBCs, but by their signaling also regulate the structure and function of the blood vessel.

Angiogenesis - by simple definition - is the formation of new blood vessels from preexisting vessels. New blood vessel formation is necessary not only for embryonic growth, but also for wound healing in adults. But angiogenesis also has its deleterious effects. During their evolution from a confined state ("carcinoma in situ") to an invasive and possibly metastatic stage, tumors must undergo an angiogenic switch to become invasive: some of the cancer cells acquire the ability to produce proangiogenic molecules. These recruit endothelial cells in the nearby blood vessels. The activated endothelial cells proliferate, migrate and form new blood vessels. This process is termed neovascularization. As a result, the tumor has access to the surrounding tissue through these new blood vessels prompting tumor progression and metastasis. Hence, the process is also referred to as tumor angiogenesis (1).

Blood vessel formation is a complex response that is driven by a plethora of molecules like chemokines, growth factors, angiopoietins, oxygen sensors, junctional proteins etc. (2). Under the influence of these driving forces endothelial cells have to grow, divide and invade the extracellular matrix. Growth factors and their receptors send signal cues to cells, which then activate the protease system to degrade the extracellular matrix. The cytoskeletal system and the adhesion molecules undergo rearrangements and redistribution facilitating the cell to move forward.

Endothelial cells receive input from many receptors that sense the composition of the extracellular milieu and then recruit intracellular signaling pathways to translate the cue into a biological response. Input from the various receptors must be integrated to allow for a coordinated response. Accordingly, cross-talk and interdependence between various pathways is expected to occur. In my thesis, I look at the interdependence of the protease system and the adhesion molecules under the influence of growth factors, which can drive the process of angiogenesis.

1.1 Vascular Endothelial Growth factor (VEGF)

A large number of mitogens capable of stimulating angiogenesis have been identified over the years; the list includes vascular endothelial growth factor (VEGF), basic fibroblast growth factor (bFGF), platelet-derived growth factor (PDGF), tumor necrosis factor-α (TNFα), angiogenin etc. Initially bFGF was considered to be the prime angiogenic factor regulating tumor angiogenesis. Several observations, however, questioned this original assumption. The mechanism that allowed for regulated secretion of bFGF was not clear and is still not understood, because bFGF lacks a signal peptide that is characteristic of secreted peptides (3). More importantly, the mitogenic action of bFGF is not restricted to endothelial cells; in fact many other cell types respond to bFGF (4). Finally, it was noted that neutralization of bFGF by antibodies did not affect tumor angiogenesis (5). Thus, a factor other than bFGF must exist to account for the capacity of tumor cells to induce blood vessel sprouting.

Vascular Endothelial Growth Factor or VEGF was first described by Napoleone Ferrara in 1989 (6). It was purified as a secreted protein from the conditioned media of bovine pituitary follicular cells based on its ability to strongly promote endothelial cell growth. The striking property of VEGF was that it was mitogenic specifically for capillary and human umbilical vein endothelial cells. At the same time *Connolly et al* described that VPF (Vascular Permeability Factor) also increases blood vessel permeability (7). VPF was later established to be identical to VEGF.

In situ hybridization analysis revealed that the expression pattern of VEGF and its receptors is correlated with embryonic angiogenesis. The transcript levels of VEGF and its receptor were abundant during embryonic brain development, but they rapidly declined in adult brain when endothelial cell proliferation had ceased (8,9). If tumors were examined, VEGF was found to be expressed by tumor cells; VEGF-receptor expression, in contrast, was confined to tumor vascular endothelial cells (10,11). Hypoxia or low oxygen tension associated with tumor necrosis also stimulates VEGF secretion by endothelial cells promoting angiogenesis in an autocrine and paracrine manner (12,13).

The mammalian VEGF family consists of 5 members: VEGF-A, B, C, D and PLGF (Placental Growth factor). These VEGF ligands can bind to three structurally similar Type III tyrosine kinase receptors: VEGFR-1 (Flt-1), VEGFR-2 (Flk-1/KDR) or VEGFR-3. VEGF-B and PLGF bind exclusively to VEGFR-1. VEGF-C, D can bind to VEGFR-2 or VEGFR-3, while VEGF-A is the most promiscuous homolog which can bind to all three receptors (14). A structurally related protein, VEGF-E is encoded by the parapoxvirus Orf virus (OV). VEGF-E has bioactive properties similar to VEGF-A but is receptor specific only for VEGFR-2 (15). VEGF-A exists in several isoforms generated by alternative splicing- $VEGF_{165}$, $VEGF_{121}$, $VEGF_{189}$, $VEGF_{206}$.

VEGFR-3 is mainly expressed on lymphatic endothelial cells and therefore mostly associated with lymphangiogenesis (16). VEGFR-1 and 2 are mainly expressed by vascular endothelial cells and considered to be key mediators of vascular angiogenesis. Both, VEGFR-2 and VEGFR-1 knockout mice die in the embryonic stage by day 8.5 (17,18). VEGFR-2-deficient mice were found to be devoid of blood vessels, mature endothelial cells and haematopoietic progenitor stem cells (17). In contrast, in VEGFR-1(-/-) mice, differentiated endothelial cells are abundant but disorganized (18). This difference in phenotype documents that the two receptors control endothelial cell proliferation, differentiation and migration in a non-redundant way; this may be accounted for by the recruitment of distinct downstream signaling pathways.

In order to understand the difference in these two receptors, porcine aortic endothelial (PAE) cells were stably transfected with either VEGFR-1 or VEGFR-2. Ligand binding prompts VEGFR activation by receptor dimerization and autophosphorylation at tyrosine residues. For VEGFA, VEGFR-1 was found to have a higher binding affinity (16 pM) than VEGFR-2 (760 pM). In spite of this low binding affinity, VEGFR-2 was the receptor responsible for most if not all actions of VEGF that can be observed: VEGF induced actin reorganization, stimulated chemotaxis and elicited a mitogenic response in VEGFR-2-transfected PAE cells. In contrast, PAE cells stably expressing VEGFR-1 did not proliferate in response to VEGF (19). The observation that VEGFR-2 is responsible for mitogenicity was also confirmed by mutating residues in $VEGF_{165}$ necessary for either VEGFR-1 or VEGFR-2 binding. $VEGF_{165}$ mutants with normal VEGFR-2 but reduced VEGFR-1 binding could elicit strong proliferative signals in bovine adrenal cortical capillary endothelial cells. However, $VEGF_{165}$ mutants with normal VEGFR-1 and reduced VEGFR-2 binding were poor mitogenic ligands (20).

The strong binding affinity but weak tyrosine kinase activity of VEGFR-1 suggested that it could be a negative regulator of angiogenesis. In fact, knock out mice which lack the intracellular domain of VEGFR-1 (VEGFR-1 $TK^{-/-}$) are perfectly healthy (150) as opposed to lethality and disorganized vasculature observed in VEGFR-$1^{-/-}$ mice (18). Therefore, the role of VEGFR-1 in angiogenesis could be to sequester and maintain a physiological balance for VEGF using its ligand binding domain and hence maintain proper blood vessel formation. A role of VEGFR-1 in inflammation has also been described. VEGFR-1 is expressed by macrophage/monocyte lineages and is necessary for differentiation and activation. VEGFR-1 $TK^{-/-}$ mice are less susceptible to chronic arthritis compared to wild type mice. In the absence of VEGFR-1 signaling several functions of macrophage are affected resulting in reduced macrophage infiltration and rheumatoid pathology (151).

Signaling via VEGFR-2 has been extensively studied. VEGF ligand binding prompts VEGFR-2 receptor dimerization and auto-phosphorylation at several tyrosine residues in the intracellular domain. Y1175 is a crucial auto-phosphorylation site as mice with a phenylalanine substitution in Y1173 (which corresponds to Y1175 in humans) die at E8.5 with similar defects as the VEGFR-2$^{-/-}$ mice (21). Auto-phosphorylation of Y1175 allows for subsequent binding of phospholipase C-γ (PLC-γ) and of several adaptor proteins such as SH2 domain-containing adaptor protein B (SHB) and Shc-like protein (SCK). Binding of PLC-γ to Y1175-P leads to MAPK/ERK1/2 activation and endothelial cell proliferation (22). The MAPK activation is mediated mainly through PLC-γ/protein kinase C (PKC) dependent activation of RAF-MEK pathway but independent of RAS. Phospholipase C-γ generates two second messengers by cleavage of membrane bound PIP_2 (phosphatidylinositol-4,5-bisphosphate) to diacylglycerol and inositoltrisphosphate (IP_3). Diacylglycerol activates the classical and the novel isoforms of protein kinase C (PKC). Inositoltriphosphate releases caclium from the endoplasmic reticulum by opening its namesake receptor, a ligand-gated channel for calcium. This supports the effective activation of classical (Ca^{2+}-dependent) protein kinase C-isoforms. The VEGF-induced, phospholipase C-γ-mediated accumulation of diacylglycerol leads to a protein kinase C-mediated phosphorylation and hence activation of RAF, the kinase upstream of MEK1 (MAP kinase kinase/activator of ERK1/2). It is somewhat surprising that the small G protein RAS is dispensable in this activation of the cascade that results in the accumulation of active dually phosphorylated ERK1/2 (extracellular signal-regulated kinase-1 and -2 = p44 and p42 mitogen-activated protein kinase/MAP kinase) (23).

Apart from PLC- γ, the adaptor protein SHB also binds to Y1175-P of VEGFR-2. However, there dose not seem to be a competition between the two adaptor proteins for the same binding site. siRNA knock down of SHB does not alter PLC-γ phosphorylation upon VEGF stimulation (24). The binding of SHB to phospho-Y1175 activates FAK/focal adhesion kinase and PI3-kinase; these kinases initiate signaling pathways that control endothelial cell migration and survival (24,25). The binding of FAK to phosphorylated SHB also leads to cytoskeletal reorganization and migration: inhibition of PI3-kinase by wortmannin which is activated downstream of FAK also prevents VEGF-A induced endothelial cell migration (26). AKT/PKB (protein kinase B) is a kinase that is activated downstream of PI3-kinase via the action of phosphoinositide-dependent kinase (PDK1). In its activated phosphorylated state, AKT phosphorylates and thereby inhibits the pro-apoptotic activity of BAD and of caspase 9; this shifts the balance in favor of cell survival (25). The cascade comprising PI3-kinase and Akt was also demonstrated to be activated only upon activation of

VEGFR-2 using ligands specific for VEGFR-1 and VEGFR-2 (26). Other phosphorylation sites such as Y1214, Y951 have also been reported to control endothelial cell migration and proliferation (14).

1.2 The plasminogen system

Endothelial cells are surrounded by a three dimensional extracellular matrix, which provides structural support as well as a heterogeneous source of biologically active molecules. The process of angiogenesis is an invasive one, in which first the basement membrane has to be breached. The basement membrane is primarily composed of laminin, type IV collagen and fibronectin. Endothelial cells must then degrade the perivascular extracellular matrix composed of proteoglycans, collagen fibers and elastic fibers. Matrix degradation is facilitated by a fine balance between proteases and protease inhibitors. The two main families of proteases involved in angiogenesis are the serine proteases of the plasminogen system and the matrix metalloproteases.

The diverse roles of the plasminogen system from embryogenesis, wound healing, tissue remodeling, angiogenesis to inflammation and cancer is well documented and appreciated. The pivotal step in the plasminogen (fibrinolytic) system is the conversion of plasminogen to plasmin. This is primarily mediated by the two serine proteases - urokinase plasminogen activator (uPA) or tissue plasminogen activator (tPA). While uPA is involved in cell migration and tissue remodeling, tPA is involved with fibrin homeostasis (27). uPA is secreted as a single chain 54 kDa pro-enzyme, referred to as pro-uPA. Pro-uPA can convert plasminogen to plasmin in trace amounts. Plasmin in turn cleaves the K158-I159 peptide bond of pro-uPA, converting it to active uPA (28). This active uPA then accelerates the plasmin generation by 200-fold (29,30). A proteolytic cascade is activated whereby plasmin that is formed degrades fibrin, vitronectin, fibronectin and activates matrix metalloproteases (MMPs) (31) (Figure 1).

Endothelial cells express MMP-1, -2, -3, -9, -14 and their inhibitors, tissue inhibitor of metalloproteinases (TIMP1 and TIMP2) (32). Activated MMP-2 degrades the basement membrane type IV collagen while, MMP-1 cleaves collagen fibrils and MMP-9 degrades proteoglycans and elastins of the perivascular ECM (33). Stimulation of human umbilical vein endothelial cells with VEGF and other growth factors like bFGF, EGF can induce the transcription factor ETS-1, which in turn induces the expression of MMP-1, -3 and -9, TIMP-1 and uPA within 2 hours (34,35). The proteolytic activity of this cascade is subject to inhibition at different levels (Fig. 2): the serpins (serine protease inhibitors) PAI-1 and PAI-2 suppress plasmin formation by inhibiting uPA and tPA, α₂-antiplasmin blocks the action of plasmin that was generated in the circulation; TIMPs and α₂-antiplasmin preclude the formation of active matrix metalloproteases.

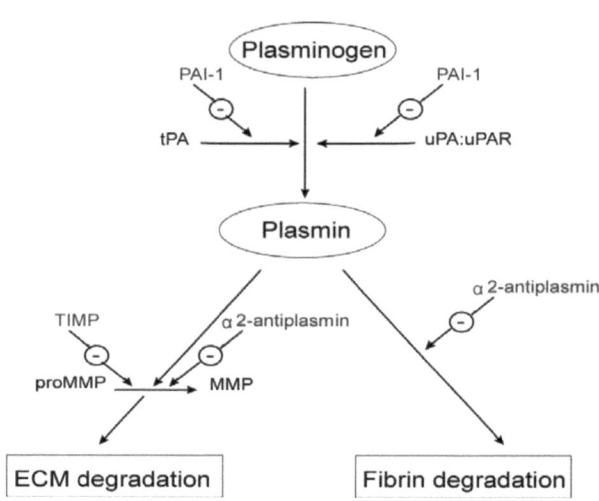

Figure 1. Schematic representation of the plasminogen system.

1.3 Structure of uPAR and of uPA

uPA is composed of three domains, the amino terminal epidermal growth factor-like domain, the kringle domain and carboxy terminal protease domain. The epidermal growth factor-like domain and the kringle domain together form the aminoterminal fragment. Both, pro-uPA and active uPA are ligands for the urokinase receptor (uPAR) with a very high binding affinity in the sub nanomolar range (K_d 0.5 nM) (36). Complex formation between pro-uPA or uPA and uPAR on the cell surface enhances plasminogen activation by about 16-fold (28). Thus, uPAR is a key player because it allows to focus plasminogen activation at a given area of the cell surface.

The urokinase receptor uPAR is also referred to as CD87; uPAR is composed of 283 amino acids and has 3 internally homologous domains of approximately 90 amino acids each, which are designated domain-1, -2 and -3 starting from the amino terminal end. The domains are separated by short linker regions. The carboxyl terminal ends with a GPI anchor, which is added to amino acid 283 during translation. Accordingly, uPAR is devoid of any intracellular domains. Site-directed photoaffinity labeling of uPAR shows that a close spatial proximity between domain-1 and domain-3 of intact uPAR is necessary for binding of uPA. The residues R53 and L66 of domain-1 and H251 of domain-3 must be brought together to form the composite ligand binding site for uPA (39). These biochemical observations were confirmed when the X-ray crystal structure of the soluble form of uPAR was solved in 2005: uPAR was found to be organized as a globular receptor with a deep central cavity (19 Å deep) and a large external surface. The amino terminal fragment of uPA occupies this cavity leaving the entire external surface open for interactions with a variety of other surface proteins; in fact, uPAR is known to form heteromultimeric complexes (40).

uPAR belongs to the Ly-6/uPAR superfamily and is heavily glycosylated with 5 potential N-linked glycosylation sites. The glycan moieties account for nearly 50% of its molecular mass of about 55-60 kDa (37). Except for N52, the other four glycosylation sites are necessary for proper folding and membrane localization of uPAR. Site directed mutagenesis of these four sites results in ER retention and no surface expression of uPAR. In contrast, if glycosylation at N52 is precluded a conformationally stable and active uPAR is expressed on the cell surface. However; due to the location of N52 in the ligand binding domain uPA binding affinity to uPAR is reduced by 4-5 fold (38).

A non-ligand binding variant of uPAR has also been identified on the cell surface. This variant lacks domain 1; uPA and plasmin can cleave cell surface-associated uPAR between domain-1 and domain-2 generating uPAR(D2+D3) (41). Cleavage by uPA can occur at two positions; between R83 and A84 and between R89 and S90 (42). Cleavage at R83/A84 exposes a chemotactic peptide SRSRY (88-92), in the linker region of uPAR. The peptide SRSRY in its soluble form is a potent chemoattractant for monocytic cells. Migration is mediated by the binding of the peptide to the G-protein coupled receptor: low affinity receptor for formyl-peptide (FPRL1), and ensuing p56/p59[hck] tyrosine phosphorylation (43,152). Cell surface bound uPAR(D2+D3) devoid of domain-1 and hence proteolytic capabilities is still capable of inducing cell migration. This is because fMLP induced chemotaxis requires the interaction of it's receptor- FPRL1 with the SRSRY peptide region of uPAR(D2+D3)

(153).

Intact uPAR may also be released from the cell surface by cleavage of the glycolipid anchor. The resulting soluble form of intact uPAR has also been found in blood sera of healthy people in very low levels (2.71±1.12 µg/l). However patients with malignant cancers have elevated levels of soluble uPAR (5.82±2.27 µg/l) in serum (154). Elevated levels of soluble uPAR are found in blood plasma and tumor tissue of many malignant cancers such as colorectal cancer (44), squamous cell lung carcinoma (45), acute myeloid leukemia (46) and breast cancer patients (47). These increased levels of soluble uPAR were found to be inversely related to patient survival; hence, cleaved soluble uPAR may be of clinical significance as a prognostic marker. The source for this cleaved soluble uPAR forms in blood plasma is most likely the tumor tissue *per se*: increased expression of uPAR was observed in metastases; these finding can be rationalized by taking into account that uPAR supports invasive growth of cells and that the metastatic burden is a prognostic determinant in many types of cancer. However, the enzymes responsible for the cleavage of soluble uPAR are not known.

1.4 uPAR in angiogenesis, proteolysis and migration

The presence of the urokinase receptor on cultured endothelial cells was first described in 1990; this discovery provided an explanation for the mechanism by which the soluble proteins uPA and tPA regulated plasminogen activation at the interface between vessel wall and blood or the subendothelium (48). A balloon catheter induced carotid injury model in rats showed very little or no mRNA expression of uPA, tPA or uPAR in the quiescent endothelium but a strong expression of all three plasminogen activator molecules *in vivo* in the regenerating endothelium at the wound edge (49). A similar expression pattern was also observed in human glioblastoma tumors *in vivo*. In situ hybridization localized uPAR mRNA to endothelial cells at sites of vascular proliferation in the leading edge of the tumor. With increasing malignancy uPAR expression also increased in the glioblastoma tumor cells. (50) It was later appreciated that blood and lymphatic vascular endothelial cells respond to specific stimulation with angiogenic growth factors like VEGF and bFGF by increased expression of uPA and uPAR, which in turn promoted plasminogen activation (51,52).

A wide variety of human cancers also express elevated levels of uPAR which is indicative of their capacity to invade tissues and to metastasize (53). This observation and the current mechanistic model of the plasminogen activation system predicts that uPAR antagonists – i.e., agents that suppress binding of uPA to uPAR and hence plasmin formation - inhibit tumor angiogenesis and metastasis formation *in vivo*. Several studies have provided a proof-of-principle by examining the action of uPAR antagonists in tumor models. An antibody that competitively inhibited uPA binding to uPAR substantially reduced bFGF-induced neovascularization and B16 melanoma growth in mice (54). Injection of PC3 prostate cancer cells stably transfected to express a catalytically inactive but receptor bound uPA lead to reduced metastasis compared to wild type PC3 cells in mice (55,56).

The generation of uPAR-deficient mice allowed for examining the function of uPAR *in vivo*. Surprisingly, homozygous uPAR$^{-/-}$ mice are fertile and viable (57). In vitro studies demonstrated a delay (3 h) in uPA mediated plasminogen activation of peritoneal macrophages isolated from uPAR$^{-/-}$ mice. This was attributed to the absence of uPAR-bound uPA on macrophage cell surfaces. uPAR$^{-/-}$ macrophages were also as efficient as wild type macrophages in matrix degradation (57). Several *in vivo* studies were carried out with uPAR$^{-/-}$ mice to examine the role of uPAR in disease models: when assessed three weeks after injury of the arterial wall, the absence of uPAR did not affect arterial neointima formation or smooth muscle cell migration (vascular wound healing) (58,59); it also did not impede fibrin clearance (60) or the formation of new blood vessels, if examined two weeks following injection of malignant keratinocytes (61). The observation of pericellular localization of uPA around uPAR$^{-/-}$ cells (58) implied that plasminogen activation could take place in a manner independent of binding to uPAR. This combined with defects in fibrin clearance (60) and neointima formation (62,63) in uPA$^{-/-}$ mice led to conclusion that uPAR seemed to have a redundant role for *in vivo* uPA-dependent plasminogen activation.

However, *in vivo*, uPAR also supports cellular migration by a mechanism that is not mediated by activation of the proteolytic cascade (*May et al.* 64). Leukocytes adhere on the vascular endothelium in a manner dependent on β2 integrins. This is an essential step in transendothelial migration. In uPAR$^{-/-}$ mice, migration of leukocytes to the inflamed peritoneum is significantly impaired compared to the wild-type mice. The impaired migration in uPAR$^{-/-}$ mice suggests that β2 integrin-dependent leukocyte migration is regulated by uPAR (64). The urokinase receptor was also found to be necessary for neutrophil recruitment in response to pulmonary infection with *Pseudomonas aeruginosa*, a gram-negative bacterium that elicits a β2 integrin-dependent inflammatory response. While neutrophil recruitment is impaired in mice deficient in uPAR, this is not seen in uPA$^{-/-}$ mice. This finding is formal proof that the recruitment of neutrophils is independent of uPA (65, 66). The modulation of β2 integrin dependent migration of leukocytes by uPAR can be rationalized by assuming that these two proteins interact directly. This conjecture was verified by immunoaffinity purification: *Bohuslav et al* used uPA as bait to isolate a large receptor complex comprising uPAR, β2 integrin (LFA-1) and phosphorylated SRC kinase (67).

While – as mentioned earlier (see section 1.3) - uPAR lacks an intracellular domain, its large external surface provides many potential binding sites. In fact, uPAR was found to interact with different integrin family members such as β1, β2 and β3 integrins, with the extracellular matrix ligand vitronectin, with GPCRs, with GP130, with mannose-6-phosphate/insulin-like growth factor receptor-2 and also with receptor tyrosine kinases like the PDGF- and the EGF-receptor. These observations supported the development of a concept in which uPAR was viewed as multifunctional receptor capable of influencing cellular events like migration, proliferation, differentiation and adhesion apart from its cognate downstream proteolytic cascade.

1.5 Integrins

Adhesion molecules mediate the anchorage of cells to the extracellular matrix. The integrin adhesion molecules are found as heterodimers, each consisting of a non-covalently linked α-subunit and β-subunit. Both subunits are type I transmembrane proteins with multiple extracellular domains and a short cytoplasmic tail. The diversity of the family is evident by the fact that there are 18 α-subunits and 8 β-subunits, which can form 25 different heterodimers. Each heterodimer binds to its own specific set of matrix ligands: in addition, multiple integrins share the same matrix ligand. By binding to matrix ligands, integrins relay information on the outside environment through their cytoplasmic tails into the cell. Many adaptor proteins like talin, vinculin and actin-binding proteins connect the cytoplasmic tail of the integrins to the cytoskeleton and to signaling molecules. Integrins exist in a latent form and are subject to activation by signals arising on the cytoplasmic side (a process referred to as "inside-out signaling"). Thus integrins are bidirectional signaling molecules, which mediate cell adhesion and support migration, proliferation, survival and differentiation.

Cell migration is contingent on the regulated association and dissociation of adhesion receptors, *i.e.*, integrins must disengage from ECM contacts at the trailing end and form new focal contacts at the leading edge (68). This requires continuous and vectorial integrin redistribution, which occurs by both, clathrin-dependent and clathrin-independent mechanisms (69). The cytoplasmic domain of β1 integrins contains two conserved NXXY-motifs responsible for recruitment of endocytotic adaptor proteins. Mutation of these motifs prevents clathrin-dependent endocytosis (70). As mentioned above, integrins undergo a transition from a latent form that does not recognize cognate ligands to a binding-competent conformation. To date it is not clear if the active and inactive forms of integrin follow the same route of endocytosis and which regulatory mechanisms define differences in endocytotic routing. This may be due to association with different surface molecules: NRP1, a co-receptor for VEGF and semaphorin 3A, controls endothelial cell adhesion to fibronectin in a manner independent of its ligands. Through its cytoplasmic association with the endocytic adaptor GIPC1, NRP-1 mediates its co- internalization with active α5β1 integrin into RAB5-positive early endosomes. Inactive α5β1 also undergoes endocytosis, albeit independently of NRP-1 (71). Migration also requires formation and disassembly of focal adhesions. In HT1080 fibrosarcoma cells, focal adhesions disassembly requires the internalization of active β1 integrins in a clathrin-dependent manner (72). The transmembrane 4 superfamily member PETA-3/CD151 colocalizes with many integrin chains such as β3, α2, α3, α5, α6 on the cell surface and in endocytic vesicles. Depletion of CD151 impedes endothelial migration and tube formation. Accordingly, CD151 has been proposed to play a major role in regulating integrin trafficking or function (73).

1.6 Integrins in angiogenesis

The state of endothelial cells (i.e., quiescent, migrating or proliferating) determines the pattern of integrin expression (see below). The activation of proteases during angiogenesis creates an enormous change to the extracellular matrix. Proteases remodel the ECM to present a promigratory environment which supports cell invasion, migration and survival. VEGF in particular increases vascular permeability (7). Thereby blood-borne provisional ECM proteins like fibrinogen, fibronectin and vitronectin gain access to the perivascular space and generate an ECM conducive to vascular remodelling. The MMP-mediated degradation of ECM is not only restricted to the clearing of the surrounding matrix; the degradation of type IV collagen and laminin exposes proangiogenic cryptic binding sites for integrins. Using cell adhesion and affinity chromatography experiments αvβ3 integrin was found to bind to the RGD sites exposed by denatured but not native type IV collagen (74). *In vivo,* MMP-2 mediates type IV collagen denaturation to expose these cryptic sites for αvβ3 integrin. These in turn promote angiogenesis and growth of tumour cells (75).

The major integrins expressed on the quiescent vascular endothelium are α1β1, α2β1, α3β1, α5β1, α6β4, α6β1, α8β1, and αvβ5 (76). The majority of these integrins are either collagen- or laminin-binding receptors, *i.e.*, components of the basal basement membrane. Integrins α5β1, α8β1, and αvβ5 are fibronectin-binding integrins. The provisional ECM is formed from the existing ECM by three processes: (i) the partial degradation of the existing ECM, (ii) the resulting exposure of cryptic binding sites in components of the existing ECM and (iii) the deposition of blood-borne soluble matrix ligands. These reactions create an ECM favourable for fibronectin- and vitronectin-binding integrins during angiogenesis and lead to increased expression of integrins α5β1, α8β1, and αvβ5 and *de novo* expression of integrin αvβ3 (77,78).

Because of their importance in supporting endothelial function, the role of integrins has been extensively studied in angiogenesis using *in vivo* as well as *in vitro* systems. Antibodies against αvβ3 and αvβ5 integrin and antagonistic peptides effectively blunt angiogenesis. Specifically, inhibition of αvβ3 integrin blunted bFGF- or TNFα-induced angiogenesis; inhibition of αvβ5 integrin suppressed VEGF-induced angiogenesis in an *in vivo* rabbit corneal or chick chorioallantoic membrane assay (80). Based on these and related findings, integrins α5β1, αvβ3/ αvβ5 are considered as candidate drug targets and their suitability for the treatment of different types of cancer is being explored in clinical trials (79). However, studies on mice, in which integrin subunits were ablated by gene targeting, cast doubt on the viability of this concept: embryonic angiogenesis is not impaired in mice rendered genetically deficient in integrin subunit β3 or β5 (81,82). In fact, tumor growth and angiogenesis are significantly enhanced upon subcutaneous injection of tumor cells in these mice. Importantly, in β3 integrin-deficient mice, VEGFR-2 expression was enhanced in endothelial cells (83). Eighty percent of the αv integrin-deficient mice die in gestation due to placental defects; the remaining 20%, which are born alive, exhibit intracerebral and intestinal hemorrhages but notably no vascular defects (84). Thus embryonic angiogenesis proceeds normally in the absence of αvβ3 and αvβ5.

The β1 integrin subfamily is essential for early angiogenesis in mice: a deletion of β1 integrin in vascular endothelial cells is embryonically lethal, the mice pups die around embryonic day E9.5-E10.5 with defects in angiogenic sprouting and vessel branching (85). Global deletion of α5 integrin is also embryonically lethal at embryonic day E10-E11: the mice pups with the homozygous deletion displayed mesodermal defects with a distended and leaky vascular system (86). Mice with conditional α5 integrin gene knock out were generated by crossing α5 integrin floxed mice with a Tie2-Cre transgene mouse line expressed in endothelial and hematopoietic cells. However, endothelial knock out of α5 caused no obvious defects in embryonic vasculogenesis or angiogenesis as the mice were viable (87) In contrast, a double knock-out of α5 and αv integrin in endothelial cells causes lethality at around embyonic day E14.5 with extensive defects in remodelling of the heart and vessels (87). The most plausible interpretation of these observation is a model in which embryonic angiogenesis does not rely on a single β1 integrin heterodimer but is rather supported by several integrins that functionally overlap. This redundancy allows for compensation in the absence of a single integrin by related integrin heterodimers or even by other proteins. However, from the perspective of drug development, it would be interesting to know how tumour angiogenesis is affected in mice in which α5-integrins is specifically ablated in the vascular endothelium.

Growth factors and receptor tyrosine kinases can control integrin recycling to regulate cell migration. Stimulation of fibroblasts with PDGF results in increased recycling of αvβ3-integrins through RAB4-positive, early endosome; this promotes cell adhesion and spreading (88). Stimulation of endothelial cells with angiopoietin-2 induces complex formation of its receptor - Tie2 with αvβ3 integrins at cell junctions. Upon this interaction αvβ3 integrin adaptor proteins – talin and p130cas - are dissociated followed by αvβ3 integrin internalization and lysosomal degradation (89). Conversely integrins can influence the recycling of other integrins and receptor tyrosine kinases: αvβ3 integrin is a negative regulator of VEGFR-2 recycling. Inhibition of αvβ3 *in vivo* leads to increased recycling of VEGFR-2 and thereby promotes angiogenesis (90). This finding also provides a plausible explanation for the aggravated tumor angiogenesis observed in β3 integrin-deficient mice (83). Thus integrin trafficking is a key modulator that can govern cell migration and signaling.

1.7 Interactions between uPAR and integrins

The consequences of uPAR-integrin interaction have been much under scrutiny since *Bohuslav et al* purified a complex containing uPAR and β2 integrin (LFA-1) from monocytes (67). *Wei et al* explored the functional significance for complex formation between integrins and uPAR by using a HEK293 cell line that overexpressed uPAR: uPAR bound to activated β1 integrins, and promoted adhesion and migration on a vitronectin matrix (96). However contradictory observations concerning uPAR and integrin interaction on a vitronectin matrix have been subsequently reported.
Vitronectin, a prototypical protein of the ECM, is a high-affinity ligand of uPAR; this was first suggested by the observation that uPA induced adhesion of anchorage-independent myeloid cell lines to purified vitronectin (91). Related findings were also observed with adherent cells: HEK293 cells over expressing GPI-anchored uPAR but not soluble uPAR strongly adhered to vitronectin in the presence of uPA (92). Using an alanine scanning library of single site uPAR mutants, the functional epitope necessary for interaction with vitronectin was found to be located outside the uPA binding cavity involving residues in domain 1 and residues in the linker peptide connecting domain 1 and 2 (93). Mutation of residues W32 and R91, which are located in domain 1 and 2, respectively, reduces vitronectin affinity five-fold but does not affect binding of uPA. HEK293 cells expressing the mutant uPAR also exhibited reduced adhesion and ERK1/2 phosphorylation when seeded on a vitronectin matrix as opposed to HEK293 cells with uPAR either intact or with mutations in integrin binding sites. Therefore the study claimed that a uPAR-vitronectin interaction is sufficient to induce adhesion and signaling in HEK293 cells (94). If microvascular endothelial cells are plated on vitronectin, its binding to both integrins and uPAR is necessary to localize uPA to the focal adhesions. Treatment of endothelial cells with an inhibitory peptide that disrupts uPAR-integrin interaction before plating on a vitronectin matrix did not affect the localization of uPA to focal adhesions (95). These studies indicate that uPAR and integrins sequentially ligate to vitronectin to mediate adhesion or uPA localization on a vitronectin matrix.

Aguirre Ghiso et al extended the notion that it is the overexpression of uPAR in tumor cells that drives the physical association of uPA, uPAR and α5β1 integrins. This coincides with increased levels of phosphorylated and hence activated FAK, SRC and ERK1/2, which support sustained tumor growth. In contrast, low levels of uPAR are associated with reduced activation of ERK1/2 and tumor dormancy (97,98). Upon testing CHO cells expressing human α or β integrin subunits for their ability to bind immobilized soluble uPAR, the interaction of uPAR with integrins was not limited to α5β1 but to also encompassed the β3 integrin family, in particular αvβ3, α3β1, α6β1 and α9β1 (99). Indirect immunofluorescence in HT1080 cells showed colocalization between uPAR and alpha integrin subunits confirming the adhesion assay studies. α5 and αv integrin colocalized with uPAR on a fibronectin matrix and α3 and α6 colocalized with uPAR on laminin. Furthermore, FRET studies verified the physical interaction between uPAR and integrins. HT1080 cells labeled with TRITC conjugated uPAR mAb and FITC-conjugated β1 or β3 mAb detected FRET values on a fibronectin, vitronectin or laminin matrix but not on polylysine. There is a critical distance for Förster resonance energy transfer (FRET): for the pair of donor flurophore FITC and acceptor flurophore TRITC, this distance is 4.9-5.5 nm. Therefore, uPAR and integrin β1 or -β3 are within 5.5 nm of each other, distances at which proteins can have direct physical contact.

Wei et al showed that soluble uPAR binds to immobilized α3β1 integrin but not α5β1 or α2β1 integrins in the presence of uPA (101). The interaction between uPAR and α3β1 integrin was demonstrated using a 17-mer integrin α3β1 inhibitor peptide (α325). UPA-mediated uPAR binding to α3β1 integrin was inhibited in the presence of α325 peptide in binding assays. In the MDA-MB-231 tumor cell line, which expresses little if any αvβ3, the addition of uPA drives complex formation between uPAR and α3β1. Concomitantly, α5β1 and α2β1 integrins are activated in a Gαi protein-dependent manner. This promotes adhesion on fibronectin and collagen. The study demonstrated that uPA and uPAR can preferentially complex with one integrin heterodimer and influence the signaling of other integrin family proteins. However a drawback is the lack of biochemical studies in the MDA-MB-231 tumor cell line ruling out possible uPA mediated uPAR and α5β1 or α2β1 integrin interaction on the cell surface (101). UPAR is overexpressed in kidney podocytes of patients suffering from progressive kidney diseases. LPS administration leads to proteinuria in wildtype but not uPAR$^{-/-}$ mice. Using a transient *in vivo* gene delivery system, which restored uPAR expression in uPAR$^{-/-}$ mice, *Changli et al* demonstrated that uPAR activates αvβ3 integrin in a uPA independent manner to promote podocyte motility and proteinuria (102). Mechanistically, activated integrin αvβ3 and uPAR complexes drive tyrosine phosphorylation of the adaptor protein p130Cas. The adaptor protein CRK binds to the phosphorylated p130Cas through its SH2 domain. Through its SH3 domain, CRK recruits Dock180- a guanine nucleotide exchange factor (GEF). Once the p130Cas-CRK-Dock180 complex is formed, Dock180 activates the Rho GTPase RAC by catalyzing the exchange of GDP for GTP. Activated RAC accumulates at the cell front promoting actin polymerization and cell migration (103,104). The work summarized above highlighted the importance of complex formation between uPAR and integrin subunits demonstrating that these complexes gave rise to intracellular signals that were translated into cellular responses. Accordingly, these insights stimulated research into the nature of the interaction site.

Wei et al also identified a peptide of 25 amino acids in a phage display library which inhibited the interaction of uPAR with both, $\beta 2$ and $\beta 1$ integrins (96,105) Using CD11b α-integrin chimeras, *Simon et al* uncovered that the exposed loop in the highly conserved non-I-domain region at the upper surface of the αM chain of $\beta 2$ integrin interacts with uPAR (105). Interestingly the peptide identified by phage display and the amino acid stretch in the αM (termed M25) chain showed sequence homology. The CD11b interacting region was also compared for their homology with sequences of other integrin α chains. Particularly the corresponding amino acid stretch of integrin α3 and α6 showed 40% sequence similarity; however, only a single amino acid residue is identical, i.e., the histidine residue corresponding to H245 (101,106). The importance of this residue is highlighted by the following findings: epithelial cells deficient in α3 integrins do not form dense clusters in cell culture. This can be restored by reintroducing α3 integrins in these cells (106). In these cells, uPAR engages α3 integrin and this results in activation of ERK1/2 via SRC and the subsequent transcriptional induction of uPA. However, this response is abrogated, if a α3 integrin harboring the point mutation H245A is expressed in α3-deficient epithelial cells (98,107).

Two separate groups worked on elucidating the uPAR interacting region with integrin. *Chaurasia et al* immobilized synthetic peptides corresponding to regions in uPAR on nitrocellulose membranes and checked for their ability to bind purified α5β1 integrin in a solid phase dot-ELISA assay. A peptide sequence of 9 contiguous amino acids in domain III of uPAR (240-248) bound to α5β1 integrin. This was followed by dot-ELISA of single site soluble uPAR mutants for the various amino acids; revealing a single amino acid – S245 whose mutation completely abrogated α5β1 integrin binding. The introduction of this single mutation in HEP3 cells as opposed to wild-type uPAR did not restore ERK activation or their growth on a chorioallantoic membrane model (108). *Wei et al* introduced point mutations for charged amino acids into the cDNA encoding uPAR and expressed the resulting mutants in HEK293 cells. These transfected cells were then probed by co-immunoprecipitation for α5β1 and α3β1 integrin binding. This approach identified the single residues H249 and D262 in domain III of uPAR capable of disrupting its interaction with α5β1 and α3β1, respectively (109). HT1080 fibrosarcoma cells and H1299 lung carcinoma cells utilize α5β1 integrins to adhere on fibronectin. These tumor cell lines were stably silenced of their endogenous uPAR and transfected with either H249 or D262 mutant uPAR. ERK and MMP-9 activation was markedly reduced in tumor cells with the H249 mutation. uPAR- $\alpha5\beta1$ integrin complexes engage with the fibronectin matrix triggering a FAK and SRC activation. SRC activates the RAC-1 pathway which leads to induction of ERK activation and MMP-9 production; disruption of the pathway leads to reduced tumor invasion (109). Similarly, H1299 cells with a (H249A-D262A) mutation in uPAR failed to induce tumor growth in an orthotopic lung tumor model (110).

Several proteins have been shown to either positively or negatively regulate the uPAR-integrin interaction and its ensuing effect on cell adhesion and migration. The mannose 6-phosphate/insulin-like growth factor 2 is considered to have tumor suppressing abilities as its absence is recorded in tumor malignancies. *Schiller et al* described a mechanism whereby mannose 6-phosphate receptor expression promotes uPAR cleavage and down regulates $\alpha v\beta3$ integrin surface expression in human kidney carcinoma cell lines. Abolition of mannose 6-phosphate by RNA interference results in increased uPAR mediated plasminogen activation, $\alpha v\beta3$ mediated cell adhesion and cell invasion. However, the molecular mechanism as to how mannose 6-phosphae regulates uPAR expression is not properly understood (111). ECRG2 - a serine protease inhibitor binds to the kringle domain of uPA without interfering with uPA and uPAR interaction. But the presence of ECRG2 in this complex impairs the interaction of uPAR with $\alpha5\beta1$ and $\alpha3\beta1$ integrin. Consequently SRC and ERK activation is down regulated with concomitant reduced cell invasion and migration (112). Tumorigenicity of human epidermoid carcinoma (Hep3) cells is driven by uPAR expression. Excess uPAR bridges the association between $\alpha5\beta1$ integrin and the receptor tyrosine kinase epidermal growth factor receptor. $\alpha5\beta1$ integrin is activated and epidermal growth factor receptor is ligand independently phosphorylated resulting in ERK activation and tumor proliferation (113).

1.8 The LDL-receptor family

LDL-Receptor gene family members are scavenger receptors which transport their bound ligand to lysosomes for degradation. Their extracellular domain consists of epidermal growth factor (EGF) precursor like repeats and complement type cysteine-rich repeats responsible for ligand binding. Depending on the type of the LDL-receptor family member the intracellular domain consists of at least one or three NPXY motifs - responsible for clathrin mediated endocytosis. LDL-Receptors bind to a variety of diverse ligands for eg; proteinases, proteinase inhibitor complexes, lipoproteins, viral proteins or tyrosine kinase receptors.

One of the members of the LDL-receptor family, LDL receptor-related protein (LRP) regulates extracellular proteolysis by mediating the catabolism of uPA-PAI-1 complexes and tPA-PAI-1 complexes. LRP is necessary for embryonic development as gene deletion in mice resulted in developmental arrest at day 13.5 (114). Inflammatory stimuli activate macrophages to migrate in a CD11b dependent manner. Macrophages require tPA, its inhibitor PAI-1 and LRP for the different steps of CD11b dependent migration on fibrin rich provisional matrix. tPA interacts with CD11b and enhances macrophage adhesion on fibrin. tPA is inactivated by PAI-1 which promotes the association of the complex of tPA, PAI-1 and CD11b with LRP. LRP mediated endocytosis of the complex switches the macrophage to a de-adhesive state promoting forward migration (115). Circulating tPA catalyzes plasmin activation which is necessary for dissolving fibrin clots in blood. Hepatic parenchymal cells of the liver are responsible for regulating tPA levels to maintain plasma fibrinolysis. This is achieved by the LRP receptor, which binds and clears tPA/PAI-1 complexes by endocytosis and lysosomal degradation (116). It has to be noted that tPA maintains fibrin homeostasis in the circulating system thereby circumventing cell anchoring for it's function. Therefore uPAR is not involved in LRP mediated clearance of tPA. Various members of the matrix metalloproteinase family MMP-13, MMP-9 and MMP-2 in complex with their inhibitors TIMP are also endocytotically cleared by LRP (114).

However, the functions of LDL-receptor family members are not limited to endocytosis but they also participate in signaling. An example of LDL-receptor mediated signaling is observed in breast cancer cell lines such as MCF-7 and MDA-MB-435 which express very low-density lipoprotein receptor (VLDLR) instead of LRP (117). Treatment of MCF-7 cells with uPA induces a transient ERK activation, which can be made sustained by the simultaneous addition of PAI-1. PAI-1 mediated sustained ERK activation in turn activates MCF-7 cell proliferation and growth. RAP – a 39 kDa protein is a high affinity ligand for VLDLR and other LDL receptor family members. The addition of RAP converts the uPA and PAI-1 mediated sustained ERK activation to a transient one; demonstrating the presence and requirement of VLDLR for PAI-1 mediated signaling (118). A similar observation was made by us in low PAI-1 expressing HT1080 fibrosarcoma cells. Addition of exogenous PAI-1 induces sustained ERK activation which can be inhibited by the addition of RAP. In addition the role of MAP kinase phosphatases was also investigated as their function is to deactivate MAPK signaling. Indeed, upon PAI-1 treatment MAP kinase phosphatases are down regulated with the concomitant ERK activation leading to increased metastatic potential (*Mihaly-Bison et al*, unpublished). These observations prove that PAI-1 can modulate uPA-uPAR signaling in cohorts with a LDL receptor family member.

LDL receptor family members can also modulate the signaling of tyrosine kinase receptors. Simultaneous stimulation of tyrosine kinases receptor like epidermal growth factor receptor, VEGF-R or FGF-R and LRP transforms the transient ERK activation of the tyrosine kinase receptor into a sustained one. However, the uPAR tri-molecular complex is necessary for sustained ERK activation as it acts as a bridging complex that links together the receptor tyrosine kinase and LRP receptor (119). LRP-1 can modulate the PDGFR-β signaling pathway by acting as a co-receptor. Activated PDGFR-β is endocytosed into early endosomes by LRP-1. Once within the endosomes, PDGFR-β kinase domain catalyzes the tyrosine phosphorylation of NPXY motif in the cytoplasmic domain of LRP-1. Phosphorylated LRP-1 is then necessary to modulate PDGFR-β mediated activation of ERK; as $LRP^{-/-}$ mouse embryonic fibroblasts (MEF) exhibit hindered ERK phosphorylation and cell growth upon PDGF stimulation (120,121).

1.9 LDL-Receptor-mediated uPAR interactions.

The principal regulator of plasminogen activation - PAI-1, binds and deactivates active uPA-uPAR complexes on the cell surface. The lack of cysteine residues makes PAI-1 unstable at $37^{\circ}C$ rendering it to convert into an inactive form. Therefore in the blood stream active PAI-1 is found bound to the somatomedin B domain (SMB) of vitronectin (123,124). Vitronectin is a major ligand for PAI-1 as it aids in its localization in the extracellular matrix and hence pericellular proteolysis. The binding region for uPAR in vitronectin is also located on the SMB domain of vitronectin (125). Both PAI-1 and uPAR share overlapping residues for their binding to vitronectin; both require the eight cysteine amino acids in the SMB domain. However uPAR in addition also requires the negatively charged Glu23 in the SMB domain; the mutation of which prevents uPAR binding to PAI-1 (125,126). PAI-1 has the upper hand in competing with uPAR for binding to vitronectin as the Kd for PAI-1/Vn is ~0.3nM and the Kd for PAI-1/uPAR is ~10nM. On a vitronectin matrix the balance between PAI-1 and uPA-uPAR govern cell adhesion and detachment. When uPA is present in excess of PAI-1, uPA governs uPAR mediated cell attachment to vitronectin. If PAI-1 is in excess it competes with uPAR for vitronectin binding prompting cell adhesion. Integrins bind to the RGD domain of vitronectin which is adjacent to the SMB domain. *Deng et al* further explained that PAI-1 hindered not only uPAR binding to vitronectin but also integrin binding to the RGD domain promoting cell detachment particularly in U937, HeLa, MCF7 and HT1080 cells (124, 126, 127).

Czekay et al provided a new light on the PAI-1 mediated cell detachment mechanism. According to them PAI-1 competes with uPAR; but rather than binding to vitronectin binds to uPA. When PAI-1 binds to the uPA, uPAR and integrin complex; integrins are inactivated and uPAR is disengaged from vitronectin. This is followed by internalization of the whole complex mediated by LDL receptor related protein (LRP) leading to cell de-adhesion. By this mechanism PAI-1 can detach cells from matrices such as fibronectin and collagen too, provided there is a high pool of uPA-uPAR-integrin complexes on the cell surface. If the ratio of matrix ligated integrins are high compared to matrix ligated integrin uPAR complexes - such as in cell lines with very little uPAR (MCF-7) - the de-adhesive effect of PAI-1 is diminished (128,129). Thus PAI-1 regulates cells adhesion independent of regulating plasmin production.

Czekay et al hypothesize that PAI-1 interaction with uPA bound uPAR inactivates and may also cause a conformational change to uPAR. This seems to trigger the deactivation of the integrin. The studies on integrin conformation changes upon uPAR interaction are limiting. uPAR on interaction with integrins modulates integrin signaling to cause cytoskeletal changes and promote cell migration (96-106). Therefore integrins interacting with uPAR alone or the uPA and uPAR complex must be in the active conformation. But in studies conducted by *Wei et al* activation specific integrin antibodies could not recognize α5β1 in complex with uPAR. *Wei et al* proposed a notion that it was difficult to envisage the activated extended conformation of α5β1 interacting with the relatively small uPAR. Therefore α5β1 adopts a modified "bent" but still active conformation not recognized by activation specific integrin antibodies on HT1080 fibrosarcoma cells (130).

The presence of the uPA-PAI-1 complex and not uPA alone initiates uPAR internalization. Mouse uPA cannot bind to human uPAR. By exploiting this species specificity *Olson et al* demonstrated that mouse LB6 cells can internalize human uPA-PAI-1 complex only if they are transfected with human uPAR. (131,132). Treatment of cells with PIPLC which cleaves all GPI-anchored proteins including uPAR or competitive treatment with amino-terminal fragment of uPA which cannot bind PAI-1 blocked internalization (131). *Nykjaer et al* demonstrated that the internalization of the tri-molecular complex is mediated by the endocytic receptor; LRP-1. As it was the addition of PAI-1 complex that triggered LRP-1 to initiate internalization, it was presumed PAI-1 would have the binding site for LRP (133,134). PAI-1 complexed with different proteinases such as high or low molecular weight uPA or trypsin was tested for LRP dependent endocytosis in MEF cells. Irrespective of the proteinase the presence of PAI-1 was necessary to initiate endocytosis. Hence PAI-1 was proposed to undergo a conformational change when bound to a proteinase and in effect reveal a high affinity binding site for LRP. Utilizing mutational studies this binding site was uncovered to locate in the heparin binding domain at R76 (135). However *Czekay et al* showed that pretreatment with purified domain III of uPAR inhibits the internalization of the tri-molecular complex; suggesting a direct binding of uPAR to LRP (136). The tri-molecular complex is internalized by LRP via clathrin coated vesicles into early endosomes (136). uPA and PAI-1 are degraded in the lysosomes and fresh uPAR is recycled to the leading edge of the cell surface (137). It is necessary to maintain this cycle of internalization and recycling in order to regulate plasmin activity, cell migration and invasion. This is so, as domain III of uPAR which hindered uPAR tri molecular complex internalization inhibited HT1080 fibrosarcoma cell migration on matrigel (136).

1.10 Rationale of the work

We have previously shown that the angiogenic response of endothelial cells to VEGF depends on the urokinase receptor. Stimulation of endothelial cells with VEGF induces rapid activation of pro-uPA bound uPAR within minutes (138). Pro-uPA activation is dependent on the VEGF-receptor-2 mediated PI3kinase signalling pathway. The PI3kinase signalling pathway is necessary for pro-MMP2 activation which in turn accelerates the conversion of pro-uPA to uPA. In order to regulate the fibrinolytic activity PAI-1 binds to the active uPA:uPAR complexes. This triggers internalization of the trimolecular complex mediated by the VLDL Receptor (131,132). Vascular cells express the VLDL Receptor and not LRP of the LDL-Receptor family (139). uPA:PAI:1 are degraded while uPAR receptor is redistributed to the focal adhesions. We have also shown that treatment with LRP abrogates VEGF induced internalization of uPAR migration on a vitronectin matrix. Importantly, treatment with RAP also significantly reduces VEGF induced *in vivo* angiogenesis on a matrigel model (139). This observation suggested that VEGF relied – at least in part – on uPAR to induce angiogenesis. Here we addressed the question, how uPAR was necessary for VEGF-induced endothelial migration on fibronectin and other typical extracellular matrix components that are not ligands for uPAR. That integrins are principal modulators of cell migration and the vast literature on integrin-uPAR interactions is already discussed here. Therefore, we initiated this work by examining for changes in integrins, uPAR-integrin interactions, and their functional influence under VEGF stimulation in endothelial cells.

2. MATERIALS AND METHODS

2.1 Reagents and Antibodies

$VEGF_{165}$ was from Promocell GmbH (Heidelberg, Germany); VEGF-E and PlGF were from RELIATech (Wolfenbüttel, Germany); recombinant mouse $VEGF_{164}$ and (2R)-2-[(4-biphenylylsulfonyl)amino]-3-phenylpropionic acid (matrix metalloprotease-2/-9 = MMP-2/MMP-9 inhibitor) from Merck (Darmstadt, Germany); wortmannin was from Sigma Chemicals (St. Louis, MO). The rat receptor-associated protein (RAP, 39kD Protein) was made as a recombinant fusion protein with glutathione S-transferase as previously described (21). The synthetic human uPAR inhibitory peptides were from Genosphere Biotechnologies (Paris, France). The sequences of the peptides are: peptide 243-251 (TASMCQHAH), Control Peptide 240-247 (GCATASMCQ) and scrambled peptide 244-252 (SCAAMHQHT). The murine uPAR- derived peptide m.P243-251 (TASWCQGSH) was purchased from piChem (Graz, Austria). Matrigel solution was from Becton-Dickinson (San Jose, CA).DIVAA kit was purchased from Trevigen (Gaithersburg, MD)

Antibodies: uPAR - 3937 monoclonal was from American Diagnostica (Greenwich, CT); the anti uPAR monoclonal antibody R2 was a kind gift of Gunilla Hoyer-Hansen (Finsen Laboratory, Copenhagen, Denmark) (41); the P4C10 mouse monoclonal antibody against $\beta1$ integrin was from Sigma (St.Louis, MO) a biotinylated version (MEM-101A) was from LifeSpan Biosciences (Seattle, WA); alternatively we also used the I2G10 from Novus Biologicals (Littleton, CO). The rat monoclonal antibody against mouse integrin subunit beta-1 was from Becton Dickinson GmbH (Heidelberg, Germany).

The antibodies P1D6 and P1B5 against the $\alpha5$ integrin and $\alpha3$ integrin subunits were also from Chemicon International; a biotinylated version against the integrin $\alpha5$ subunit (X-6) was from Acris (Rancho Santa Margarita, CA). The Clathrin heavy chain rabbit monoclonal antibody and rabbit polyclonal antibody against P-FAK (Tyr397) was from Cell Signaling Technology (Beverly,MA). The rabbit polyclonal antibody against Beta3 was from Millipore (Billerica, MA) and FAK rabbit polyclonal antibody was from Calbiochem (Darmstadt, Germany). The biotinylated anti mouse antibody from Sigma (St.Louis, MO) and rat monoclonal antibody against mouse CD31 antibody from Dianova (Hamburg, Germany)

Rabbit antiserum 7 raised against amino acids 8 to 23 of the G protein $\beta1/\beta2$-subunits was used as loading control for surface biotinylation experiments (141). Alexa Fluor 488 or 568-conjugated anti-mouse or anti-rabbit IgG antibodies and the Zenon One Mouse IgG_1 Labeling Kit were purchased from Molecular Probes (Leiden, The Netherlands). Alexa Fluor 488 conjugated Fab-fragments (Zenon kit) were from Molecular Probes (Leiden, The Netherlands); Streptavidin 647 from Molecular Probes (Leiden, The Netherlands). Goat serum was from (Dako Diagnostics, Vienna, Austria).

M199 (Sigma-Aldrich, St. Louis, MO), Bovine endothelial cell growth supplement (Technoclone, Vienna, Austria), Dulbecco modified Eagle medium (Invitrogen, Carlsbad, CA), heparin (PromoCell; Heidelberg, Germany), Triton X-100 (Sigma, MO), sulfo-NHS-LC-biotin (Pierce, Rockford, IL), Complete™ protease inhibitors, Roche Diagnostics GmbH, Vienna, Austria), streptavidin-coated Sepharose 4B beads (Zymed Laboratories, San Francisco, CA), Femto-ECL reagents for enhanced chemiluminescence from Pierce (Rockford, IL), Adhesive silicone masks from (Secure Seal, Grace Biolabs, Germany), Transwell™ inserts from Corning LifeScienes (Lowell, MA).

2.2 Cell Culture

HUVECs (human umbilical vein endothelial cells) of up to passage 5 were used for all experiments. The cultures were maintained in M199 (Sigma M5017) supplemented with 20% FBS, 3 mg/ml ECGS and 22.5 mg/ml heparin (PromoCell C-30140, Germany). Cells were maintained at at 37° C with 5% CO_2. Experiments were performed using subconfluent cultures, which were rendered quiescent by incubation for 24 hr in medium containing 5% FCS, followed by serum-deprivation for additional 4 h in M199 containing 1% BSA.

2.3 Cytofluorimetric Analysis

Monolayers of subconfluent endothelial cells were treated as indicated in the various experiments with growth factors ($VEGF_{165}$, VEGF-E) for the appropriate time at 37°C. Following stimulation, the cells were transferred to ice, washed once with ice-cold PBS, and harvested with 3 mM EDTA. The harvested cells were fixed with 4% paraformaldehyde (Sigma, MO) for 15 min, and – in those instances where surface-bound and total levels of integrins were assessed - aliquots were permeabilizied with 0.2% Triton X-100 (Sigma,MO) for 10 min. Non-specific binding sites were blocked with 2% goat serum (Dako) followed by incubation with the appropriate primary antibody for 60 min at room temperature. After a PBS wash to remove unbound antibodies, the samples were incubated with an Alexa Fluor 488-conjugated (H+L) secondary antibody (Invitrogen, CA), washed and analyzed with FACSort (Becton-Dickinson). For the cell surface α integrin screening, cells were treated as above, and stained using primary antibodies from alpha integrin IHC kit. In the case of experiments using inhibitors or peptides, the cells were pretreated with the appropriate amounts for 10 min prior to stimulation. For the PI-PLC treatment, endothelial cells were treated either with 5U ml^{-1} PI-PLC at 37°C for 15 min or with buffer as control. After treatment the supernatants were removed and cells were gently washed with M199 containing 10 mg ml^{-1} BSA.

2.4 Surface Biotinylation and Western Blotting

For surface biotinylation experiments, cells were grown in 100 cm^2 dishes and serum starved as above. Either before or after stimulation with VEGF, cells were washed three times with ice-cold PBS (pH 8) and incubated with 2 mM Sulfo-NHS-LC-Biotin (Pierce, IL) for 30 min on ice to biotinylate surface proteins. The reaction was quenched by the addition of an ice-cold solution containing 200 mM glycine in PBS (pH 7.0) to consume excess biotinylation reagent followed by two PBS washes. Subsequently, the cells were scraped off in 1 ml of prechilled PBS containing a cocktail of protease inhibitors (Complete, Roche), and briefly spun at 4°C. The pellets were then lysed in RIPA Buffer containing 0.1% sodium dodecyl sulfate (SDS), 1% Triton X-100, 1% deoxycholate, 0.15M NaCl and 0.1M Tris-HCl ph 7.6 and protease inhibitors for 30 min on ice. The cell extracts were centrifuged at 13,000 rpm for 10 min at 4°C, and clear supernatants were collected. The biotinylated surface proteins were adsorbed from the lysates by overnight incubation with Sepharose 4B Streptavidin beads (Zymed Laboratories, CA). The beads were washed three times with RIPA buffer and the adsorbed material eluted by boiling in non-reducing sample buffer. Proteins were separated using SDS polyacrylamide gels, transferred to PVDF membrane and probed with relevant antibodies–mouse monoclonal (R2) for uPAR and mouse monoclonal (I2G10) for integrin β1 subunit. Immunoreactive bands were detected by enhanced chemiluminescence according to manufacturer's instructions (Pierce, IL).

For analysis of uPAR cleavage experiments, endothelial cells were seeded at a density of 9×10^5 in 6 well plates and grown to confluence. Following serum starvation as above, samples were stimulated with either VEGF$_{165}$ (50ng/ml) or PMA (100nM) for the indicated time periods. In some cases, protease inhibitor cocktail (Complete, Roche) was also added. Following stimulation, samples were transferred to ice and washed once with cold PBS. Cells were lysed using lysis buffer containing 4% SDS and 10% β-mercaptoethanol. ~20µg protein per sample was separated using SDS polyacrylamide gels and probed for uPAR as described above.

2.5 siRNA treatment of endothelial cells

The predesigned uPAR siRNA and the scrambled siRNA was from Ambion (LifeTech, Austria). Endothelial cells were transfected with siRNA at a concentration of 100nM using X-tremeGENE siRNA transfection reagent from Roche, (Diagnostics GmbH, Vienna, Austria) in serum free medium according the manufacturer's instructions. 36 hours after transfection, cells were starved and subjected for FACS analysis as above.

2.6 Immunocytochemisrty

Serum-starved endothelial cells seeded on gelatin-coated cover slips were stimulated as described. Following stimulation with VEGF, cells were washed with PBS, fixed with 4% paraformaldehyde for 15 min, permeabilized with 0.1% Triton-X for 10 min and blocked with 5% goat serum in primary antibody dilution buffer (DAKO) for 60 min at room temperature. The samples were incubated with the appropriate primary antibodies in 2% goat serum overnight at 4°C. After washing, cells were incubated for 45 min with secondary antibodies conjugated to appropriate fluorophores. Following washes, samples were mounted in Vectashield (Vector Laboratories; Burlingame, CA). The primary antibodies used were mouse anti-human uPAR antibody (3937) and rat anti-human β1 integrin antibody (Mab13). Secondary antibodies were goat anti-mouse Alexa Flour 568 and goat anti-rat Alexa Flour 488 (1:250) respectively. The images were captured using a Zeiss Axiovert 200 LSM 500 microscope at 60-fold magnification. Colocalization was quantified after image thresholding and background substraction using the colocalization threshold plugin of ImageJ program. Twenty to 25 individual cells were analyzed.

2.7 Micropatterning

The micropatterning experiment was done as described previously by Schwarzenbacher et al. (142) with modifications. Poly (dimethylsiloxane) stamps containing the microarray (square of size and depth 3 µm) were generated by standard photolithography. Stamps were rinsed with 100% ethanol and distilled H_2O, dried under as tream of N_2, and incubated with 100 mg/ml Cy5 labelled BSA for 30 min at room temperature. The stamps were then extensively washed with PBS and distilled H_2O and dried under N_2. The stamps were then placed under their own weight onto epoxy-derivatized glass cover slips for 30 min. Upon removing the stamps, the cover lips were sealed with adhesive silicone masks (Secure Seal, Grace Biolabs, Germany). The Cy5-labelled BSA micropatterned glass cover slips were then incubated for 1 h at room temperature with 50 µg/ml streptavidin dissolved in PBS, rinsed with PBS and finally incubated with 10 µg/ml biotinylated monoclonal antibody. BSA efficiently blocks unspecific adsorption of both streptavidin and mAB, thereby providing a well-defined 3 µm micro pattern. Serum-starved endothelial cells were incubated with or without VEGF$_{165}$ were seeded on micropatterns and allowed to adhere. Following stimulation for 60 min, the cells were fixed with paraformaldehyde at a final concentration of 4%. The samples were stained for uPAR with the monoclonal mouse 3937 antibody (prelabeled with Alexa Fluor 488-conjugated Fab-fragments (Zenon kit, Molecular Probes) for 60 min. Images were acquired using an epi-fluorescence microscope (Axiovert 200/M, Zeiss, Germany) in the total internal reflection (TIR) configuration using an oil immersion objective (100-fold magnification).

Images were analyzed using a program that allowed for semi-automated image and data analysis that accomplished the following tasks: an automatic gridding algorithm calculated the grid-size and the rotation angle φ of the image. The grid subdivided the total image into adjacent squares, which were quantified according to the average specific signal with a central circle (F^+) and unspecific background outside the circle. (F^-). This information was used to calculate fluorescence (F) and contrast (C).

2.8 Transwell Migration Assay

Transwell inserts (8 µm pore, Costar, Inc.) were coated with 5 µg/ml fibronectin for 1 h and were blocked with 2% bovine serum albumin in phosphate-buffered saline for 1 h. Serum starved endothelial cells (25,000) were added to the upper chamber with or without the inhibitory peptides (25 µM) or RAP (40µg/ml) and allowed to migrate for 4 h at 37 °C. The cells that had not migrated were then removed from the upper chamber by wiping the upper surface with an absorbent tip. Cells that had migrated to the lower side of the Transwell insert were then fixed for 15 min with 4% paraformaldehyde and stained with DAPI 1:1000 (Pierce) for 15 min. After extensive PBS washing, the number of cells that had migrated was counted by using fluorescence microscopy with a 10× objective of an Olympus AX70 microscope. The assay was performed in triplicates as data is expressed as percentage of migration.

2.9 DIVAA *in vivo* angiogenesis assay

For the DIVAA *in vivo* angiogenesis assay, angioreactors – (silicon cylinders of 1.0 cm length and 0.15 cm diameter) were filled with basement membrane extract either alone or containing 2 µg/ml VEGF$_{164}$ and 1mg/ml heparin with or without 250µM of the mouse inhibitory peptide at 4°C. 6-8 week old BL6 mice were anaesthetized with ketamin (100mg/kg) and xylazin (10mg/kg). A reactor each was subcutaneously implanted on either sides of the flank of the mice. The reactors were excised out on the 11th day after sacrificing the mice and frozen in liqN$_2$. Sections were stained with rat anti-mouse CD31 antibody and Hoechst. Tissue samples were visualized with an AX-70 Olympus microscope, photographed using an OPtronics DEI-750D CCD camera and analysed using the Cell$^{P®}$ imaging software (Olympus).

2.10 Statistics

Statistically significant differences were assessed by using Student's t test for paired observations. Two or more groups were compared with the control group by ANOVA followed by Dunnett's test for multiple comparisons as post hoc test.

3. RESULTS

3.1 Internalization of uPAR and β1 integrin upon VEGF stimulation.

We first examined VEGF induced changes of β1 integrins in endothelial cells. We focused on the β1 subfamily of integrins, because members of this sub-family are considered to be the adhesion molecules utilized in the unstimulated resting state of endothelial cells. Resting endothelial cells were stimulated with $VEGF_{165}$ for 30 min and immunostained for the β1 integrin subunit. In resting endothelial cells β1 integrin subunit was found mainly at the cell periphery: the immunoreactivity outlined the cell borders (Figure 2A, control). In contrast, in $VEGF_{165}$-stimulated endothelial cells immunostaining at the cell borders declined and increased in the vicinity of the nuclei (Fig.2A, VEGF). This shift suggested that VEGF caused internalization of integrins. We verified this by two methods: flow cytofluorometric analysis and cell surface biotinylation.

Serum starved endothelial cells were stimulated with or without $VEGF_{165}$ and analyzed for surface expression by flow cytofluorometric analysis. The surface expression of integrin β1 subunit decreased significantly by ~40% following stimulation with $VEGF_{165}$ for 60 min (Figure 2B). uPAR internalization was confirmed in parallel. uPAR surface expression also decreased by ~35% as reported previously by *Prager et al* (140). $VEGF_{165}$ dose response curve for integrin β1 subunit internalization shows maximum response at 50 ng/ml (Figure 2C). The biotinylated surface proteins were enriched by affinity purification on streptavidin-coated beads and immunoblotted for both β1 integrin subunit and uPAR. When surface biotinylation was carried out prior to stimulation with VEGF (Figure 2D), the amount of biotinylated receptors recovered from the whole cell lysates did not differ between stimulated or unstimulated cells. This indicates that proteins are not degraded or cleaved from the cell surface during the stimulation time period. However, if biotinylation of surface proteins was done after VEGF stimulation the stimulated cells had approximately ~60% and ~50% of biotinylated β1 integrin and uPAR respectively on their cell surface compared to the unstimulated cells. This reflected the reduction in surface receptor levels of uPAR and of β1 integrins upon $VEGF_{165}$ stimulation of endothelial cells observed before by flow cytometry.

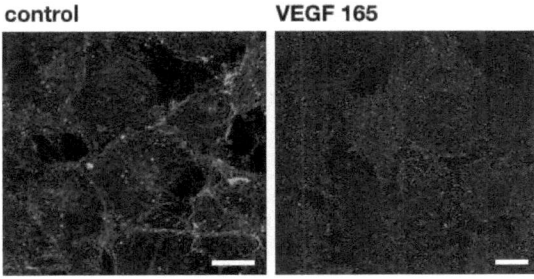

Figure 2A

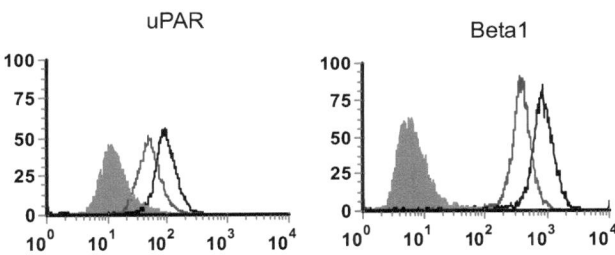

Figure 2B

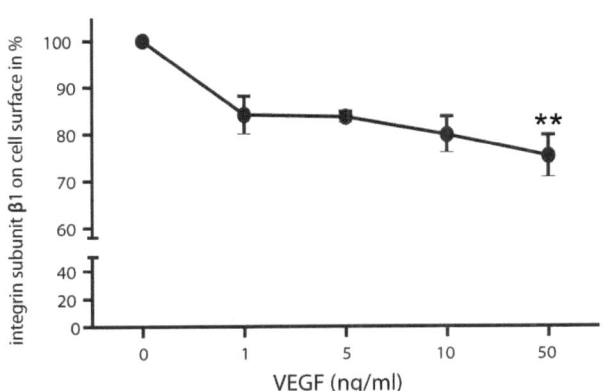

Figure 2C

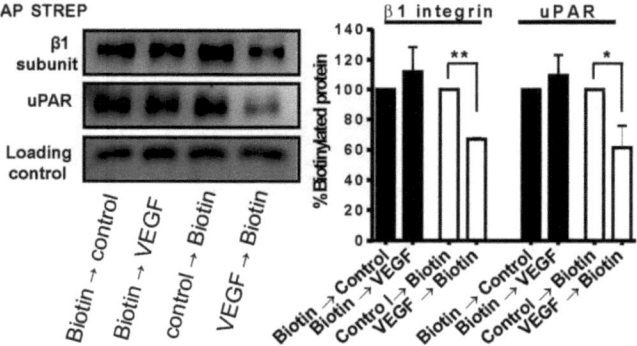

Figure 2D

Figure2. VEGF stimulation induces internalization of β1 integrins in endothelial cells. (A) Serum starved endothelial cells were stimulated for 30 min with $VEGF_{165}$ (50ng/ml). Samples were fixed, permeablized and stained for integrin β1 subunit (green). Confocal image were taken at 60x magnification. scale bar 10 µm (B) $VEGF_{165}$ induces internalization of uPAR and β1 integrins. Representative immunocytofluorimetric histograms of endothelial cells, stimulated with or without $VEGF_{165}$ (50ng/ml) are presented. Upon stimulation under serum starved conditions for 60 min at 37°C, the cells were transferred to ice and harvested with EDTA. The samples were fixed and analyzed for integrin β1 subunit or uPAR surface expression. VEGF stimulated sample-red histogram, Control sample-black histogram, No staining-grey histogram. (C) Dose response curve for VEGF-induced integrin subunit-β1 internalization. Endothelial cells were stimulated with $VEGF_{165}$ (1-50 ng/ml) for 60 min. Samples were analyzed for integrin subunit β1 surface expression by flow cytometry as above. Line graph summarizes data from 3 experiments. Mean ± SEM, **$p< 0.01$ (D) Cell surface proteins were biotinylated with 2 mM Sulfo-NHS-LC-Biotin either before or after $VEGF_{165}$ stimulation for 60 min at 37°C. The labeled surface proteins were enriched by streptavidin affinity purification from the lysates and probed for the integrin β1 subunit (12G10 antibody) and uPAR (R2 antibody) by Western blotting. Loading control: G protein β-subunit. Blots from a typical experiment are shown (n=3). The percentage of biotinylated surface protein on stimulated cells compared to the unstimulated control was quantified from the densitometric values for both uPAR and β1-integrin. Mean ± SD, **$p< 0.01$

3.2 uPAR is not cleaved during short term VEGF stimulation.

VEGF induces activation of metalloproteases that can cause cleavage of surface molecules (34,35). Proteases and enzymes can either cleave the intact uPAR or only the Domain 1 leaving behind uPAR(D2+D3) (41,44). We checked if stimulation of endothelial cells with VEGF could induce cleavage or shedding. Endothelial cells were stimulated with either $VEGF_{165}$ or phorbol-12-myristate-13-acetate (PMA) for different time periods, and analyzed for total uPAR content by western blot. By using an antibody that detects both forms of uPAR, we observed that intact uPAR levels did not decline but modestly increased over several hours (12 h) in VEGF-stimulated endothelial cells. In the presence of the PMA (phorbol-12-myristate-13-acetate), which was used as positive control, D2D3 fragments accumulated already after 6 h In contrast, VEGF-induced cleavage was modest after 6 h and increased up to 36 h (Figure 3A). We also checked if this increase in uPAR(D2+D3) fragments is due to the activity of proteases present in the culture medium. It was so; as pre-treatment with a cocktail of protease inhibitors inhibited the formation of uPAR(D2+D3) fragments (Figure 3B). Thus, short term-stimulation with VEGF (60 min) caused a decline in cell surface levels of both, uPAR and β1 integrins that was not accounted for by proteolytic degradation but by internalization.

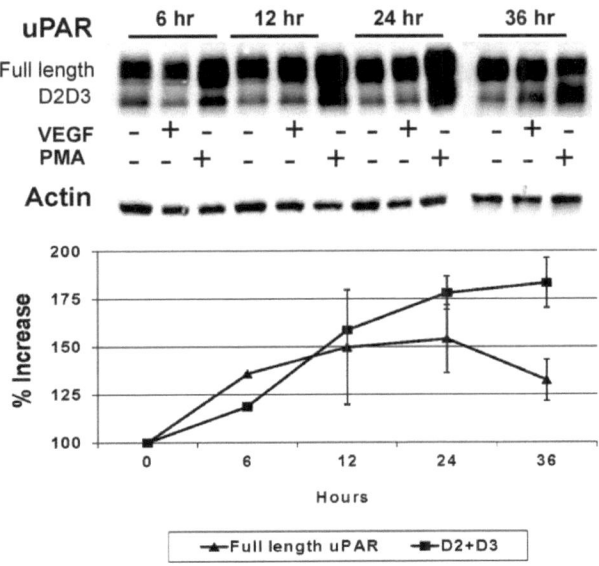

Figure 3A

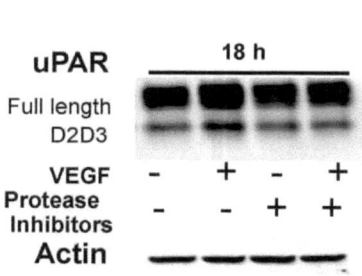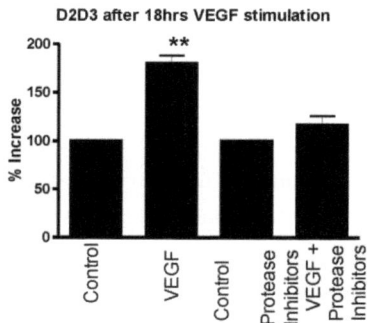

Figure 3B

Figure 3. uPAR is not cleaved during short term VEGF stimulation. (A) Serum starved endothelial cells were stimulated with either VEGF (50ng/ml) or PMA (100 nM) for the indicated time periods. Thereafter, the cells were lysed and probed for uPAR using the R2 antibody which detects both the full length and cleaved forms of uPAR by western blotting. Loading control: Actin. Blots from a typical experiment are shown (n=3). Graphs are plotted from the densitometric values after normalization with actin (ratio of uPAR level to actin; as compared to the control). Mean ± SE. **(B) Cleavage of uPAR due to long term VEGF stimulation is protease dependent.** Serum starved endothelial cells were stimulated with VEGF (50ng/ml) with or without protease inhibitors for18 h. Thereafter, the cells were lysed and subjected to western blot as above. (n=3) Mean ± SD, **$p< 0.01$

3.3 uPAR and β1 integrin interact upon VEGF stimulation.

With both uPAR and β1 integrin internalizing at a similar time interval upon VEGF stimulation, we explored for a possible interaction between these two proteins at the cell surface. We used the technique of micropatterning, which was devised to assess protein-protein interactions at the cell surface, as an independent approach to demonstrate the interaction between uPAR and integrins in

VEGF-stimulated endothelial cells (142). In this technique grids of BSA-Cy5 are printed on functionalized glass coverslips, and the interspaces are filled with sterptavidin and biotinylated antibody. We used a non functional biotinylated antibody to the integrin subunit β1. Accordingly, β1 integrins on the attaching and spreading cells, were captured at these spots of immobilized antibody. After fixation, the samples were immuno-stained for uPAR. In unstimulated cells uPAR was distributed homogeneously with occasional punctuate staining (Figure 4Aa). In contrast, upon VEGF-stimulation of endothelial cells, uPAR was redistributed to the pattern defined by the immobilized integrin antibody (Figure 4Ac). This patterning documents the interaction between the two cell surface receptors in living cells. The quantitative analysis for the single spot fluorescence yielded a mean contrast of 0.35 for stimulated cells (Figure 4Ad) and a mean contrast 0.07 for unstimulated cells (Figure 4Ac). This higher mean contrast between areas of "antibody-spots" and the background "BSA-grid" provides a quantitative readout for the extent to which uPAR became enriched in areas of immobilized integrins. In contrast, if an activating antibody against integrin-subunit β1 was used as bait, we did not observe any association between uPAR and β1 integrins upon VEGF stimulation (Fig. 4B). These observations are consistent with our hypothesis that the uPAR tri molecular complex interacts with inactive integrins upon VEGF stimulation.

We also tested if PECAM1 (CD31) and VCAM1 (CD106) - two adhesion molecules show redistribution patterns on a β1 integrin micropattern surface. Both PECAM1 and VCAM1 have no reported interaction with integrin β1 (Figure 4C,D). Endothelial cells grown on β1 integrin micropattern surfaces were stimulated with VEGF, fixed and immunostained for either PECAM1 or VCAM1. No difference was observed in the distribution pattern of either of these two proteins between non-stimulated or stimulated cells.

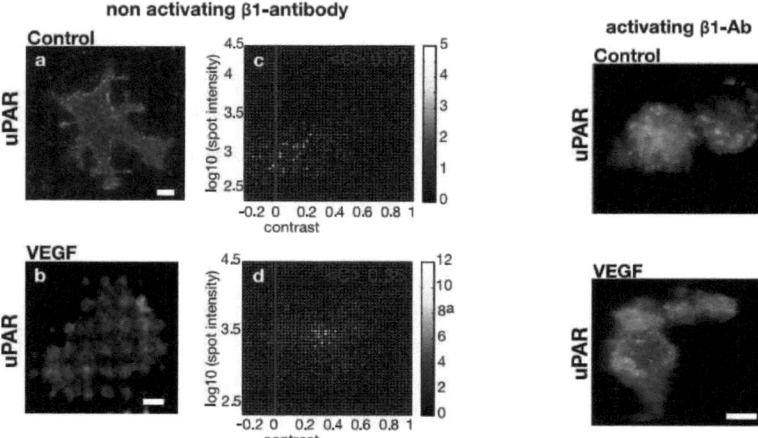

Figure 4A

Figure 4B

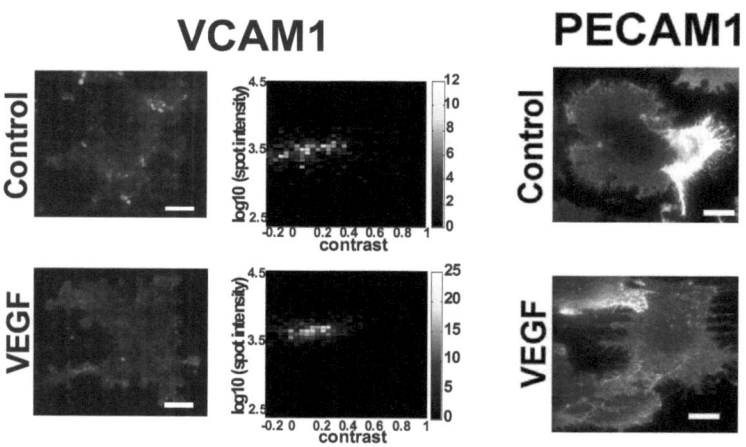

Figure 4C

Figure 4D

Figure 4. VEGF-induced interaction of uPAR and β1 integrins. (A) Cell surface interaction between uPAR and β1 integrins assessed by micropatterning. Endothelial cells were grown on functionalized glass coverslips, with grids of BSA-Cy5 printed on them and interspaces filled with streptavidin and biotinylated monoclonal antibody against integrin subunit β1. Upon stimulation with VEGF$_{165}$ for 60 min, samples were fixed and immuno-stained with uPAR monoclonal antibody 3937 pre-labeled with Zenon Alexa Fluor 488 Fab fragments. Representative TIRF images were taken using an Axiovert 200/M microscope (100x objective). UPAR is seen redistributed to the β1 antibody pattern, indicating interaction upon VEGF stimulation. Statistical analysis for the single spot fluorescence contrast C *vs.* brightness F of multiple cells (n= 64) is represented in a color density plot (b,d).. **(B)** Endothelial cells were grown on an activating β1 integrin antibody patterned surface and treated as in (A). No interaction was observed with uPAR upon VEGF stimulation. **(C&D)** Micropatterning analysis on glass coverslips patterned with either monoclonal antibody against VCAM1 (C) or PECAM1 (D). Following stimulation of endothelial cells seeded on the patterns with VEGF$_{165}$ (50ng/ml) for 60 min, samples were fixed, immunostained for uPAR and TIRF images taken as above. Scale bar 6 μm

3.4 uPAR and β1 integrin co-internalize in endocytic vesicles upon VEGF stimulation.

Based on the above observations, we surmised that uPAR and integrins co-internalize upon VEGF-stimulation. We employed confocal immunostaining to verify it was so. In unstimulated resting cells, uPAR is distributed homogenously. Upon VEGF$_{165}$ stimulation uPAR is redistributed and concentrated to newly formed focal adhesions (Figure 5A). VEGF stimulated endothelial cells were next immunostained for uPAR, integrin subunit β1 and β3. In unstimulated endothelial cells β1 integrin was mainly found at the cell-matrix border, β3 in the perinuclear compartment and uPAR was distributed in a distinct, but diffuse pattern (Figure 5A, control). Treatment with VEGF$_{165}$ led to the redistribution of uPAR to the newly formed focal adhesions where it co-localized with β3 integrins (Figure 5B, VEGF, pink arrows). Interestingly, uPAR was also found together with β1 integrins in vesicle-like structures within the cytoplasmic compartment (Figure 5B, yellow arrows). This observation confirmed their co-internalization into the same endocytotic vesicles. Integrin and uPAR redistribution was not observed, if endothelial cells were incubated with RAP prior to stimulation with VEGF (Figure 5B, VEGF/RAP). This indicated that an LDL-receptor like molecule mediated the co-internalization of uPAR and β1 integrin. We also confirmed the above observations in polarized cells. A scratch wound was made on a confluent resting monolayer of endothelial cells. Cells were allowed to migrate into the scratch overnight thus becoming polarized. Following VEGF stimulation samples were immuno-stained for uPAR, β1 and β3. The VEGF effect seen in polarized cells were reminiscent of resting endothelial cells. Focal adhesions positive for either uPAR or β3 integrin was observed in the unstimulated polarized cells. uPAR and β1 integrin colocalized in vesicle like structures (Figure 5C, white arrows) upon VEGF stimulation. Pretreatment with RAP inhibits the uPAR - β1 integrin colocalization. Integrins in the active state convey intracellular signals through the adaptor proteins like FAK. The phosphorylation state of the associated FAK can convey the activation state of the integrin. Therefore we immunostained endothelial cells stimulated with or without VEGF for uPAR, pFAK and integrin β1 subunit. In unstimulated cells pFAK colocalised with β1 integrin in point contacts (Figure 5D, blue arrows) and in occasional focal adhesions (Figure 5D, blue arrow head). Thus β1 integrins are the principal integrins which forms focal contacts in the resting state. Upon VEGF-stimulation colocalization of β1 integrins and phosphorylated FAK was reduced. VEGF promoted co-localization of uPAR and β1 integrins in vesicular structures (Figure 5D, orange arrows) and uPAR-pFAK colocalization in focal adhesions (Figure 5D, pink arrows). The reduction of phosphorylated FAK associated with β1 integrin verifies our hypothesis that VEGF stimulation inactivates β1 integrin; which leads to its internalization coupled with uPAR.

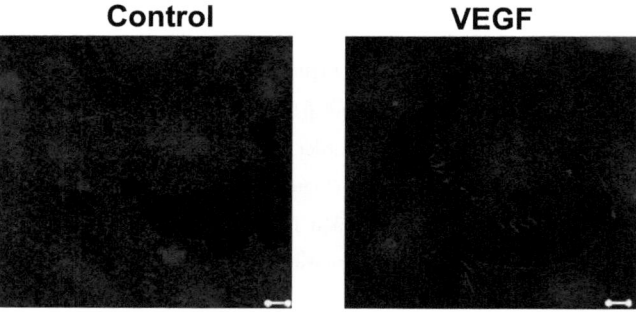

Figure 5A

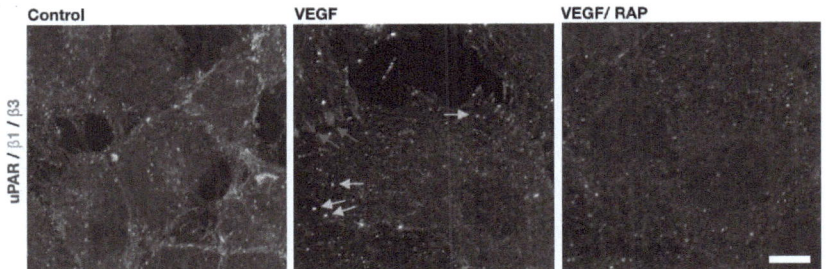

Figure 5B

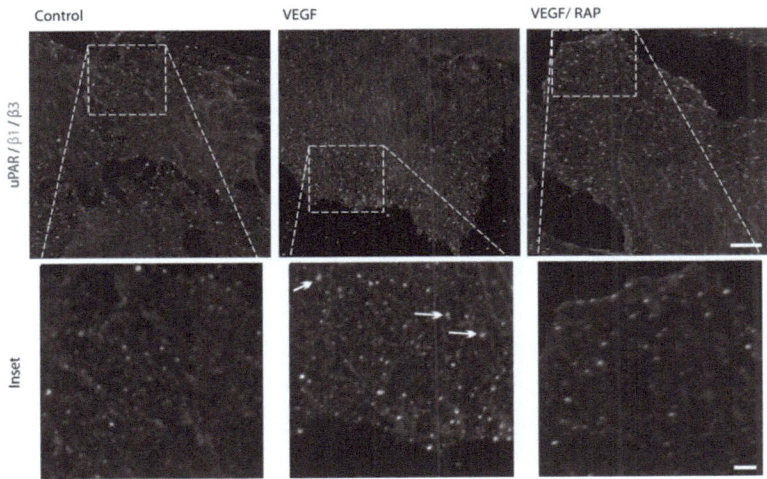

Figure 5C

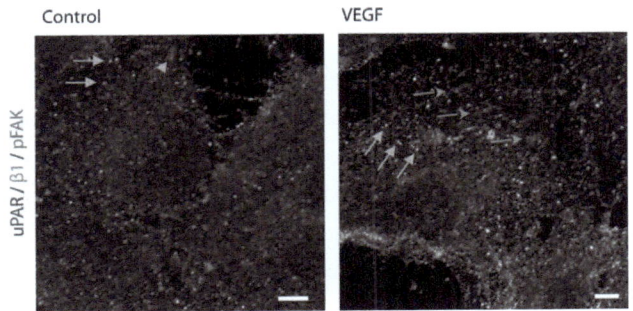

Figure 5D

Figure 5. (A) uPAR is redistributed to focal adhesions upon VEGF stimulation. Serum starved endothelial cells were stimulated with $VEGF_{165}$ (50ng/ml) for 60 min. Samples were fixed with 4% paraformaldehyde for 15 min and permeabilized with 0.1% Triton-X for 10 min. Following blocking with 5% goat serum for 60 min, samples were immunostained for uPAR (red) and confocal images taken using LSM-510 microscope at 100x magnification. Scale Bar 10 µm **(B)** Serum starved endothelial cells were simulated with $VEGF_{165}$ (50ng/ml) for 60 min fixed and permeabilized. Sample were immunostained for uPAR (red) and β1 integrin (green) and β3 integrin (blue). Scale Bar 5 µm. Upon VEGF stimulation uPAR and β1 integrin colocalize in vesicle like structures (yellow arrows), while β3 integrin and uPAR colocalize in focal adhesions (pink arrows). Pretreatment with RAP prevents the $VEGF_{165}$ induced colocalization of uPAR and integrins. Scale Bar 5µm **(C)** Polarized endothelial cells were stimulated with $VEGF_{165}$ (50ng/ml) for 60 min and treated as in (B). uPAR and β1 integrin are colocalized upon VEGF stimulation (white arrows). Scale Bars 10 and 2,5 µm **(D)** Endothelial cells were simulated with $VEGF_{165}$ (50ng/ml) for 60 min and immunostained for uPAR (red) and β1 integrin (green) and pFAK (blue). Unstimulated cells show colocalization of β1 integrin subunit and pFAK in focal adhesions and contacts (blue arrows). After VEGF stimulation, β1 integrin and uPAR co-localize in vesicles (yellow arrow), while uPAR was also found in focal adhesions co-localizing with pFAK (pink arrow). scale bar 10 µm

3.5 VEGF induced PI3-kinase pathway modulates β1 integrin internalization.

We had reported that uPAR internalization upon VEGF stimulation was induced specifically through VEGF Receptor-2 (140). We verified if β1 integrin also responded similarly to VEGF receptor-2 stimulation. Serum starved endothelial cells were stimulated with VEGF-E, a ligand specific for VEGF receptor-2 or $VEGF_{165}$ a ligand which binds to all three VEGF Receptors. Analysis for cell surface expression by flow cytometry shows that following stimulation for 60 min, β1 integrin showed reduced surface expression compared to the unstimulated cells, for VEGF-E (Figure 6A). This implies that β1 integrin internalization responds to the same internalization signals as uPAR. Stimulation of VEGF receptor-2 triggers the coordinated activation of several signaling pathways (e.g., RAS-dependent signaling, phospholipase C-γ mediated protein kinase C activation, stimulation of SRC, etc) (22-26). These support the concerted reprogramming of endothelial cells that gives rise to angiogenesis. Our previous experiments showed that VEGF induced uPAR internalization via a signaling pathway comprising VEGF receptor-2, PI3-kinase and matrix metalloproteinase (MMP-2) (140). Accordingly, we assessed the effects of the PI3-kinase inhibitor wortmannin and of the MMP-2/9 inhibitor (2R)-2-[(4-biphenylylsulfonyl)amino]-3-phenylpropionic acid on integrin internalization. Serum starved endothelial cells were pretreated with these inhibitors and stimulated with VEGF-E, the specific ligand for VEGFR-2. FACS analysis showed that VEGF-E induced β1 internalization was also blunted upon inhibition of PI3-kinase and MMP-2/9 inhibitor (Figure 6Bb,c). We interrogated the signaling network by using additional inhibitors, i.e., PD98050 for blocking activation of ERK1/2, U73122 as an inhibitor of PLC-γ and PP1 to suppress the non-receptor tyrosine kinase SRC. The rationale for examining the RAS-RAF-MEK1-ERK cascade was our recent observations that activation of a receptor tyrosine kinase and of LRP-1 in the presence of uPAR resulted in sustained ERK activation and increased cellular adhesion (119). It was therefore conceivable that the VEGF-R2 and uPAR cooperated to control integrin internalization via the ERK cascade. This was not the case: VEGF-stimulated internalization of endothelial uPAR or integrins was not inhibited by the MEK1-inhibitor PD98059 (Figure 8Bd). We also examined, whether input via PLC-γ or SRC affected recycling of VEGF-induced uPAR or integrin recycling, because PLC-γ consumes the substrate of PI3-kinase and because SRC plays a prominent role in signals arising from focal adhesions. However, VEGF-induced internalization of uPAR or of integrins was neither inhibited by the SRC inhibitor PP1 nor by the PLC inhibitor U73122. Taken together, our results show that β1 integrin internalization is controlled by the same pathway (i.e., PI3-kinase-dependent activation of MMP-2) as uPAR.

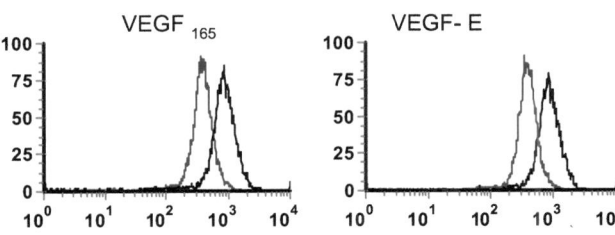

Figure 6A

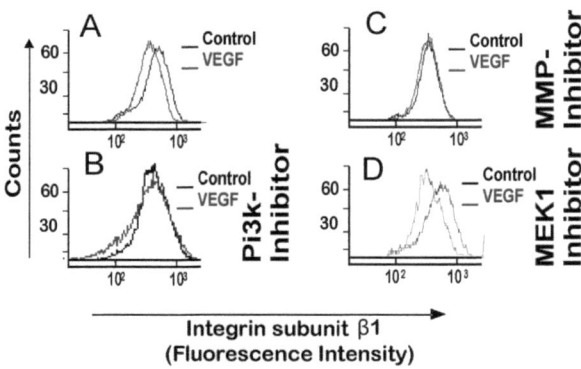

Figure 6B

Figure 6. VEGF induced β1 internalization requires VEGFR-2 dependent PI3-kinase/MMP-2 activation. (A) Serum starved endothelial cells were stimulated with either $VEGF_{165}$ (50ng/ml) or VEGF-E (50ng/nl) for 60 min, fixed and analyzed for integrin subunit β1 surface expression using flow cytometry as in Figure 5A. VEGF stimulated sample-red histogram, Control sample-black histogram.**(B) VEGF-E induced β1 internalization is inhibited by MMP and PI3-kinase inhibitor.** Serum starved endothelial cells were stimulated with VEGF-E (50ng/ml) for 60 min. Prior to stimulation, cells were treated with either MMP-2/9 inhibitor {(2R)-2-[(4-Biphenylylsulfonyl)amino]-3-phenylpropionic acid} (1μmol/L), PI3-kinase inhibitor – wortmannin (100 nmol/L) or MEK1 inhibitor - PD98059 (30μmol/L) for 30 min. Cells were fixed and analyzed for integrin subunit β1 surface expression using flow cytometry. The experiment was repeated thrice, and representative histograms are shown.

3.6 Domain 3 of uPAR is necessary for β1 integrin endocytosis.

The three domains of uPAR fold to form a urokinase binding pocket with a large external surface accessible to other ligands (40). Different amino acid residues in domain-3 (DIII) have been implicated to interact with the integrin β1 subunit; particularly S245 (108) and H249 (109). Inspection of the uPAR 3D structure shows that these residues are located side by side on the external surface opposite the urokinase binding pocket.

We synthesized peptide 243-251 which contains both candidate residues to define the uPAR interacting region required for integrin internalization upon VEGF induction. We treated serum starved endothelial cells with Peptide 243-251 and with its scrambled version prior to stimulation with $VEGF_{165}$ and determined surface levels of β1 integrin and of uPAR by FACS. As positive control we used peptide 240-248, used by (*Chaurasia et al, 2006*) to demonstrate uPAR/integrin β1 interaction. Peptide 243-251 and the positive control peptide inhibited $VEGF_{165}$-induced internalization of β1 integrin (Figure 7A). In contrast, the scrambled peptide did not have any inhibitory effect. The inhibitory peptide did not affect the VEGF-induced internalization of uPAR *per se* (Figure 7B). This indicates that the requirements for co-internalization were asymmetrical: internalization of β1 integrin was contingent on complex formation with uPAR, but VEGF-driven endocytosis of uPAR did not require an interaction with β1 integrin.

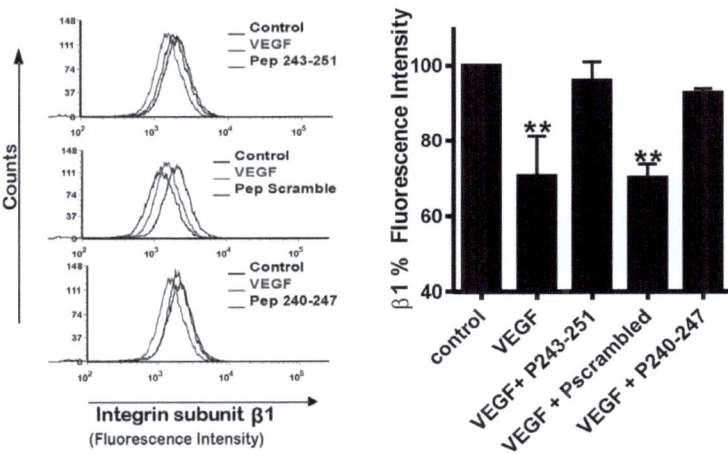

Figure 7A

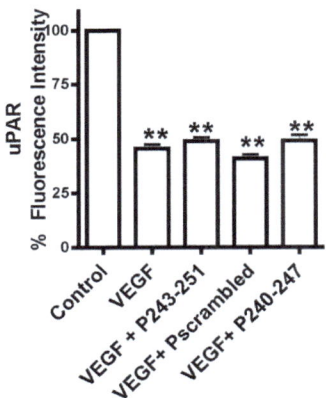

Figure 7B

Figure 7. Peptides derived from Domain-3 of uPAR block VEGF-induced integrin internalization. (A) Representative histograms show cell surface integrin β1 expression of endothelial cells pretreated with 25 µM of the inhibitory peptide (P243-252) or it's scrambled version for 10 min and stimulated with $VEGF_{165}$ (50ng/ml) for 60 min. P240-247 was used as a positive control. Right hand panel shows the quantification of percentage surface expression compared to control; calculated from the geometric mean fluorescence values of three independent experiments. Mean ± SD, **$p < 0.01$. **(B)** Quantification of uPAR surface expression after treatment as above **$p < 0.01$.

3.7 Integrin α5β1 associates with uPAR for VEGF induced endocytosis.

Six different members of the β1 family have been described on endothelial cells (76,78). Hence, we sought to determine which β1 integrin heterodimers specifically undergo VEGF induced internalization. We analyzed VEGF-induced change in surface expression of different α integrin subunits by flow cytofluorometric analysis. These experiments revealed that α3β1 and α5β1 integrins were internalized upon $VEGF_{165}$ stimulation while the surface expression of αv integrin subunit was increased. The latter can be presumed to reflect increased cell-surface translocation from the peri-nuclear vesicles (Figure 8A).

We next checked whether both α3β1 and α5β1 integrin interact with uPAR upon VEGF simulation. We verified this by resorting again to the micro-patterning approach. Micro-biochip surfaces were functionalized with an antibody to either integrin subunit α3 or α5. This forced the integrins α3β1 or α5β1 on adhering cells to accumulate according to the pattern of the immobilized antibody. We then analyzed the localization of uPAR before and after $VEGF_{165}$-stimulation of the cells. UPAR accumulated in the spots of immobilized α5β1 integrin upon VEGF stimulation (Figure 8B). This can only be accounted for by an interaction between the two receptors. In stimulated cells, a high mean contrast of 0.21 was found between the areas of antibody-immobilized integrins versus BSA-treated areas. In contrast, the mean contrast (0.04) was low in unstimulated cells. This confirms the interaction of uPAR and α5β1 integrin. The patterned integrin α3β1 antibody did not cause any patterning of uPAR upon VEGF stimulation (Figure 8C). This indicates that these two proteins do not interact to an appreciable extent in living endothelial cells.

Immunocytochemistry shows that stimulation of quiescent endothelial cells by VEGF results in a marked redistribution of α5β1 integrins. In resting endothelial cells integrin subunit α5β1 was found mainly at the cell periphery. In contrast, in $VEGF_{165}$-stimulated endothelial cells, immunostaining at the cell borders declined and increased in the vicinity of the nuclei as punctate structures (Figure 8D). This shift also confirms VEGF induced internalization of α5β1 integrins.

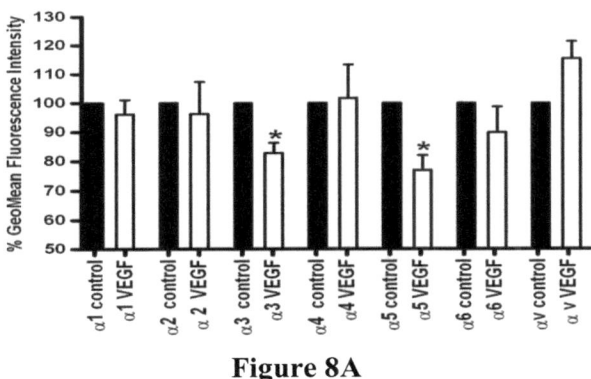

Figure 8A

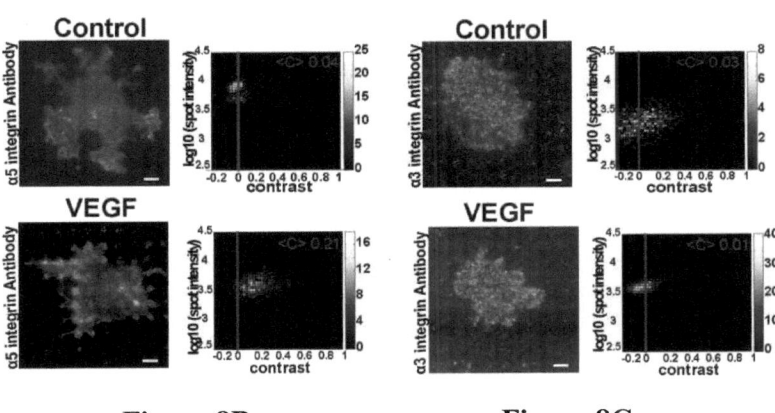

Figure 8B   **Figure 8C**

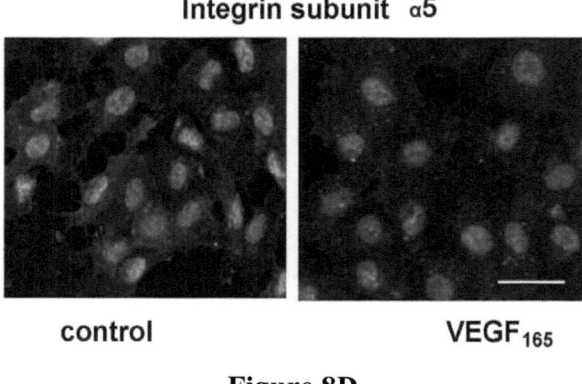

Figure 8D

Figure 8 The α5β1 integrin heterodimer interacts with uPAR upon VEGF stimulation and undergoes internalization. **(A)** Alpha integrin subunit screening for VEGF induced internalization in endothelial cells. Serum starved endothelial cells were stimulated with or without VEGF$_{165}$ (50ng/ml) for 60 min, fixed and analyzed for surface expression level of different α-integrin subunits by flow cytometry. Representative bar diagram shows percentage cell surface α-integrin expression compared to it's respective control taken as 100%; calculated from the geometric mean fluorescence values of four independent experiments. Mean ± SD, *p< 0.05 **(B&C)** VEGF$_{165}$ stimulation of endothelial cells induces co-distribution of α5β1 and uPAR on a micro-patterned surface. Endothelial cells were grown on microbiochips patterned with an antibody to the α5 **(B)** or α3 **(C)** integrin subunit. After stimulation with VEGF$_{165}$ for 60 min, samples were fixed and immuno-stained using a pre-labeled uPAR monoclonal antibody. Representative TIRF images show uPAR redistribution to α5β1 integrin but not α3β1 antibody patterns upon stimulation. Statistical analysis of multiple cells was done as in Figure 7A and depicted in color density plots. 100-120 cells were evaluated per condition. Scale bar 6μm. **(D)** VEGF$_{165}$ stimulation reduces α5-integrin (green) at the cell periphery. Resting endothelial cells were stimulated with VEGF$_{165}$ (50ng/ml) for 30 min, fixed and immuno-stained for integrin α5 subunit. Images were taken using LSM-50 microscope with a 40x objective.

3.8 uPAR is essential for VEGF induced internalization of α5β1 but not α3β1.

We verified if for the observed internalization of α5β1 integrins, the concomitant internalization of uPAR is co-incidental or required. We employed 3 different strategies:- a) phosphatidylinositol-specific phospholipase C (PI-PLC) treatment b) Pretreatment with uPAR inhibitory peptide and c) transient knock down of uPAR. Endothelial cells were pretreated with PI-PLC to cleave off all GPI-anchored proteins including uPAR from the cell surface. Surface expression of α5β1 and α3β1 integrins was subsequently assessed by FACS analysis after VEGF$_{165}$ stimulation. Addition of VEGF$_{165}$ led to internalization of ~25% and ~35% of α3β1 and α5β1 integrin, respectively. However, the lack of uPAR impaired VEGF induced α5β1 integrin but not α3β1 integrin internalization (Figure 9A versus Figure 9B). Pretreatment of endothelial cells with the uPAR inhibitory peptides also yielded similar results. The inhibitory peptide P243-251 and the control peptide P240-248 inhibited the internalization of α5β1 but not that of α3β1 integrin (Figure 9 C,D). The scrambled peptide did not have any inhibitory effect. A dose response curve for the inhibitory effect of P243-251 and the scrambled peptide was done. P243-251 could inhibit α5β1 integrin internalization starting from a concentration of 25µM with no obvious effect of the scrambled peptide (Figure 9E). We also knocked down uPAR in endothelial cells using siRNA strategy. Transfection of endothelial cells with uPAR siRNA strongly inhibited the expression of uPAR compared to the scrambled siRNA (Figure 9F – western blot). Upon stimulation with VEGF uPAR silenced endothelial cells did not show α5β1 integrin internalization (figure 9F). These observations indicate that the complex formation between uPAR and α5β1 integrin is necessary to drive efficient endocytosis of α5β1 integrin.

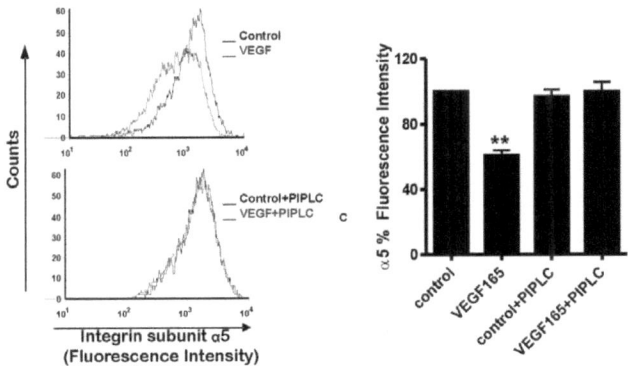

Figure 9A

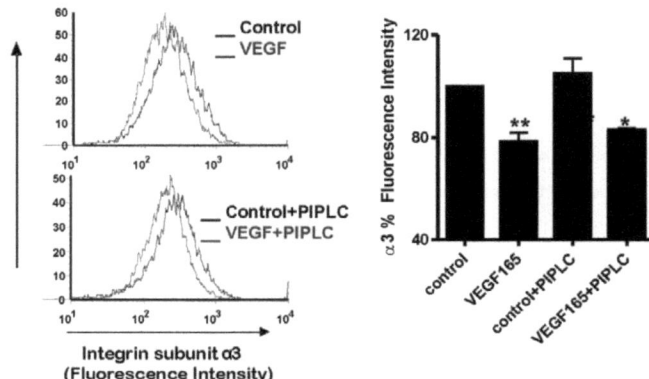

Figure 9B

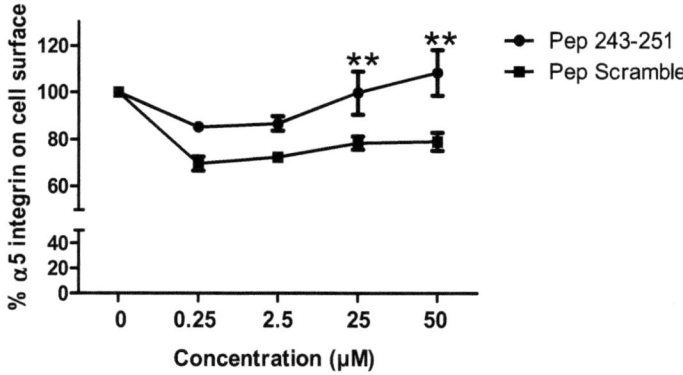

Figure 9C

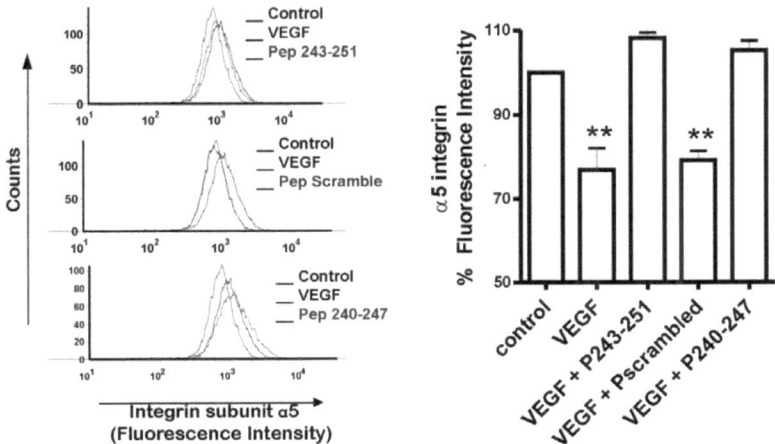

Figure 9D

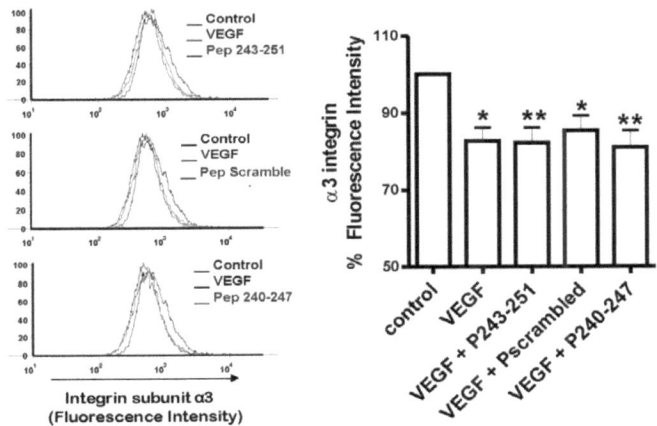

Figure 9E

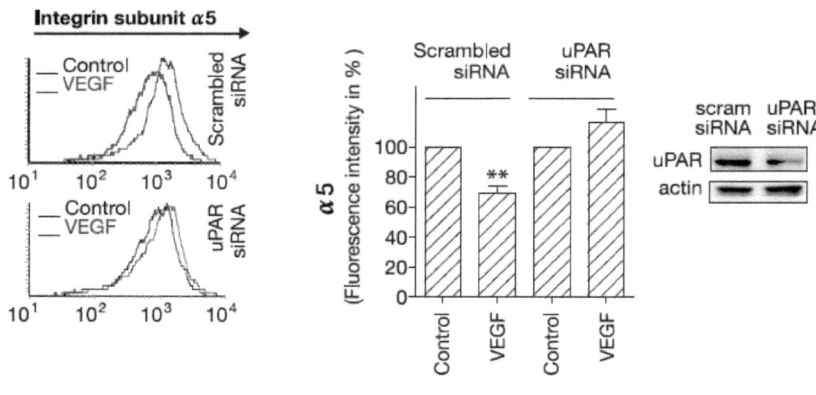

Figure 9F

Figure 9 uPAR is necessary for VEGF induced internalization of α5 integrin subunit but not α3. (A) Cleavage of uPAR from endothelial cell surface diminishes VEGF induced α5-integrin internalization. Serum starved endothelial cells were treated with 5U ml^{-1} of PI-PLC for 15 min, stimulated with VEGF$_{165}$ (50ng/ml) for 60 min and prepared for flow cytometry as described in Figure 5A. Quantitative analysis of cell surface α5 integrin subunit calculated from the geometric mean fluorescence values of four independent experiments is shown in the right panel. Mean ± SD, **$p < 0.01$, *$p < 0.05$ **(B)** The experiment was repeated as above but analyzed for α3 integrin surface expression. (n=3) **(C)** Dose response of P243-251 inhibitory effect of VEGF induced α5β1 integrin internalization. Endothelial cells were pre-treated with P243-251 or scrambled peptide with concentrations ranging from (0.25µM-50µM). Samples were stimulated with VEGF$_{165}$ (50ng/ml) for 60 min and assessed for α5β1 integrin internalization as above. Complete inhibition of α5β1 integrin internalization was observed at a concentration of 25µM. n=4 Mean ± SEM, **p< 0.01 **(D)** Representative histograms show cell surface expression of α5-integrin on endothelial cells pretreated with the uPAR-derived (P243-251, P240-247; 25 µM) or the scrambled control peptide (PScramble, 25 µM) for 10 min and stimulated with VEGF$_{165}$ (50 ng/ml) for 60 min. Right panel shows bar diagram summarizing 4 experiments (means±S.D.) with immuno-staining in control cells representing 100%. **(E)** Conditions were as in panel A, but surface levels of α3-integrin were quantified. **(F)** Knock-down of uPAR by siRNA impairs VEGF-induced internalization of integrin α5β1. Endothelial cells were transfected with either uPAR or scrambled siRNA (100nM) for 36 h. Knock-down of uPAR was assessed by immunoblotting for uPAR and actin (loading control) in lysates of siRNA-trasfected cells. Samples were stimulated with VEGF$_{165}$ (50ng/ml) for 60 min and assessed for α5β1 integrin internalization as above. n=3 Mean ± SD, **p< 0.01

3.9 The LDL-Receptor is necessary for VEGF induced endocytosis of α5β1 integrin.

Endocytosis of the uPAR tri-molecular complex is mediated by the LDL-Receptor (134-137). We have shown that VEGF induced endocytosis of uPAR can be inhibited by RAP - the high affinity ligand of LDL-Receptor (140). As α5β1 integrin requires uPAR interaction for its internalization, it would be predictable that VEGF induced endocytosis of α5β1 integrin can be inhibited by RAP. This was the case as pre-treatment of endothelial cells with RAP inhibited the internalization of α5β1 integrin (Figure 10A). As expected, this was not the case for α3β1 integrin (Figure 10B). Thus hindrance of the LDL-Receptor like protein inhibits both uPAR and α5β1 integrin internalization. The uPAR inhibitory peptides inhibited the internalization α5β1 but not uPAR (Figure 7A). This suggests that the complex for internalization assembles in hierarchical order and that the internalization of α5β1 is contingent on uPAR. Taken together, these data indicate that it is specifically α5β1 integrin that interacts with uPAR upon VEGF stimulation. α3β1 integrin internalization occurs independently of uPAR mediated by a separate VEGF dependent pathway.

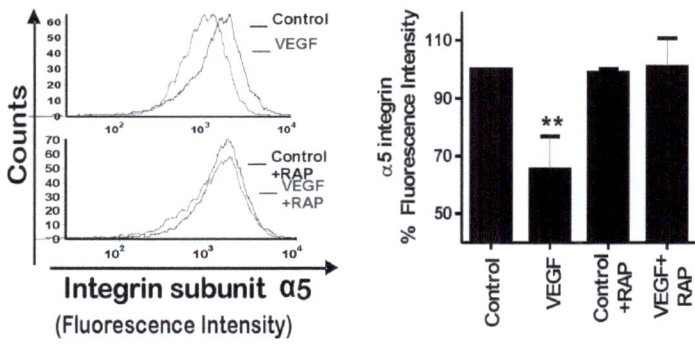

Figure 10A

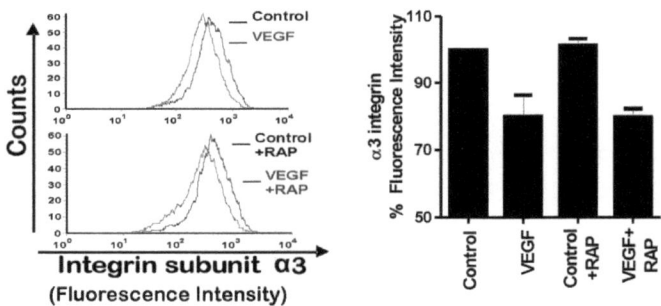

Figure 10B

Figure 10 VEGF-induced α5β1-integrin internalization requires LDLR-protein dependent uPAR interaction. Endothelial cells were treated with RAP (200 nM) for 30 min prior to stimulation with VEGF for 60 min; surface expression of α5 integrin **(A)** and α3 integrin **(B)** expression by flow cytofluorometry. Bar diagrams summarize data from 3 experiments (means ± S.D.). Statistical significance was assessed by ANOVA (**$p < 0.01$).

3.10 Inhibition of uPAR-α5β1 interaction impairs VEGF induced migration of endothelial cells.

Our results till now indicate that uPAR specifically interacts with α5β1 integrin in response to VEGF stimulation. α5β1 integrin is a fibronectin receptor and fibronectin is an essential component of the provisional extracellular matrix generated during wound healing and angiogenesis, (77,78) but is not a ligand for the urokinase receptor. If co-internalization of integrin and uPAR is functionally relevant, its disruption is predicted to have an effect on VEGF-induced endothelial cell migration. This was examined in a modified boyden chamber assay where the filter inserts were coated with fibronectin. Serum starved endothelial cells treated with the uPAR-derived inhibitory peptide were allowed to migrate towards a VEGF gradient. The inhibitory peptide significantly reduced migration towards VEGF (Figure 11A). This observation is consistent with the interpretation that uPAR-mediated integrin internalization is a necessary step in VEGF induced cell migration.

We also examined whether RAP (which precluded LDL-receptor like protein mediated uPAR-integrin internalization) suppressed VEGF induced migration of endothelial cells on fibronectin and collagen. We tested on collagen because collagen degradation exposes RGD binding sites conducive for α5β1 integrin ligation (74). In a modified boyden chamber chemotaxis assay RAP (Figure 11B,C) significantly reduced migration towards VEGF in both fibronectin as well as a collagen matrix. Taken together our results demonstrate that efficient integrin recycling is contingent on the co-internalization of uPAR, integrins and LDL-receptor like protein. Blockage of uPAR or LDLR-like protein impairs integrin recycling and in turn cell migration.

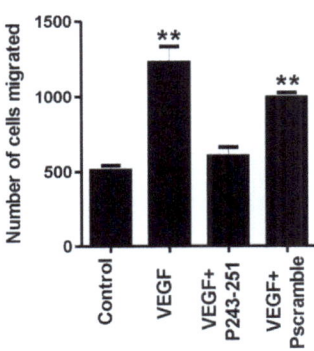

Figure 11A

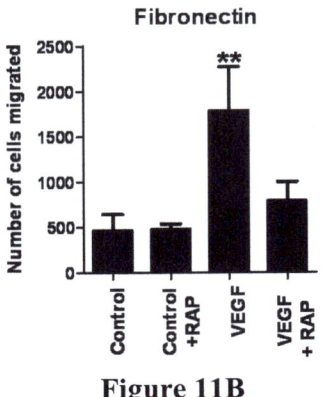

Figure 11B

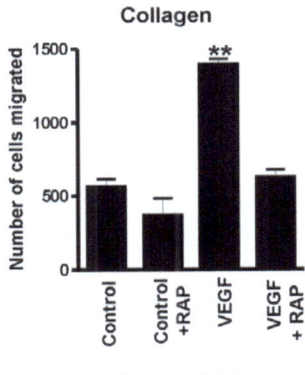

Figure 11C

Figure 11 VEGF-induced endothelial migration in vitro requires the interaction of the fibronectin receptor α5β1 integrin with uPAR. Serum-starved endothelial cells were treated with the uPAR-derived (P243-251, 25 µM) or the control peptide (PScramble, 25 µM) **(A)** or with RAP (200 nM) **(B&C)** for 10 min and then allowed to migrate through transwell membranes towards $VEGF_{165}$ for 4 h at 37°C. The transwell inserts were coated with fibronectin (5µg/ml) **(A,B)** or collagen (50µg/ml) **(C)** overnight at 4°C prior to seeding of the cells. Cells that had migrated to the underside of the membrane were fixed and counted (AnalySiS® software, Olympus). The results from three **(A)** and four **(B,C)** independent experiments are shown; statistically significant differences were verified by ANOVA (**$p < 0.01$).

3.11 UPAR- integrin interaction is necessary for *in vivo* angiogenesis.

Based on our observations inhibition of uPAR-mediated integrin recycling must have an effect on endothelial cell migration *in vivo*. We investigated the anti-angiogenic effect of the mouse uPAR-derived peptide m.P243-251 in a directed *in vivo* angiogenesis assay (DIVAA). The peptide m.P243-251 "TASWCQGSH" corresponds to the domain 3 peptide of human uPAR. Angioreactors filled with basement membrane extract containing either mVEGF or the combination of mVEGF and m.P243-251 were subcutaneously implanted for 11 days in the dorsal flanks of wild type BL6 mice. As predicted, capillary tubes formed in reactors containing mVEGF. In contrast, angiogenesis was significantly inhibited in reactors filled with VEGF and the inhibitory peptide m.P243-251 (Figure 12A). The presence of endothelial cells was verified by immuno-staining sections from the reactors for CD31 (Figure 12B,C). Sections from mVEGF containing angioreactors exhibited well developed blood vessels confirming the macroscopic examination of the reactors. Compared to the mVEGF, reactors containing mVEGF and the inhibitory peptide showed little or no CD31 positive staining. These results prove that the uPAR mediated integrin recycling is a necessary prerequisite for *in vivo* angiogenesis. In the absence of efficient recycling of integrins endothelial cells cannot invade and reorganize to form functional blood vessels.

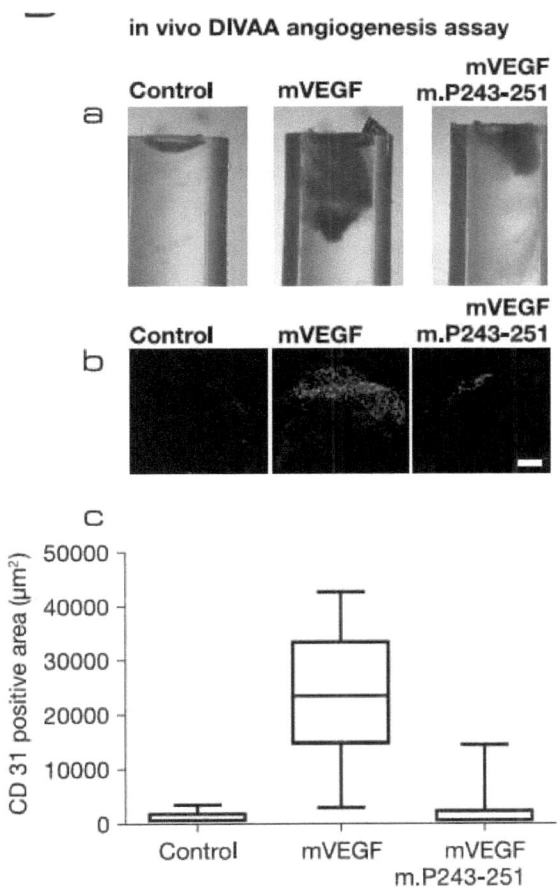

Figure 12

Figure 12 uPAR mediated integrin recycling is necessary for in vivo angiogenesis. Angioreactors of size 1.0 cm length and 0.15 cm diameter were filled with 25 µl basement membrane extract either alone or containing 2 µg/ml $VEGF_{164}$ and 1mg/ml heparin with or without 250µM m.P243-251. A reactor each was implanted on either side of the dorsal flank of 6-8 week BL6 mice for 11 days. Eight reactors were used for each group. (A) Representative angioreactor from each group retrieved after 11 days. (B) The liquid N_2 frozen angioreactors were sectioned and stained for rat anti mouse CD31 (green) and Hoescht (blue). Immunofluorescence pictures were taken using the AX-70 Olympus microscope. scale bar 100 µm (C) Sections were analyzed and quantified for invading endothelial cells (CD31 positive area) using the CellP® imaging software (Olympus). n=8 $p<0.05$; ANOVA

4. DISCUSSION

VEGF-directed reprogramming shifts endothelial cells from quiescence to an activated state that enables invasion of the surrounding tissue. Invasion is supported by the redistribution of uPAR to the leading edge of endothelial cells, which results in focused proteolysis of the extracellular matrix (52,140). Invading endothelial cells sample the environment for guiding cues by dynamically changing membrane protrusion. Because these advancing cellular filopodia are continuously remodeled, uPAR must also be subject to continuous redistribution. This is achieved by complex formation between the GPI-anchored uPAR and a member of the LDL receptor family and their subsequent internalization (133-136). VEGF drives this endosomal recycling of uPAR by a signaling cascade that emanates from the VEGF-receptor-2/Flk1 (140). Our current observations document that the VEGF-signal is funneled through uPAR to control a second, integrin-dependent limb of the angiogenic response. VEGF-induced recycling of α5β1 integrin requires uPAR and is contingent on the internalization of a complex that also contains LDL-receptor related protein (Illustrated in Figure 15). This conclusion is based on the following findings: (i) VEGF promoted assembly of a complex comprising uPAR and α5β1 integrin. (ii) This complex was internalized upon VEGF stimulation; accordingly β1 integrins and uPAR were visualized in the same intracellular compartment. (iii) The internalization of uPAR and α5β1 integrin is dependent on the LDL-receptor like protein. (iv) In the absence of uPAR, VEGF failed to trigger internalization of α5β1 integrin and thus to initiate the redistributive cycle of integrin endocytosis and exocytosis. This translated into impaired endothelial cell migration *in vitro* and *in vivo*.

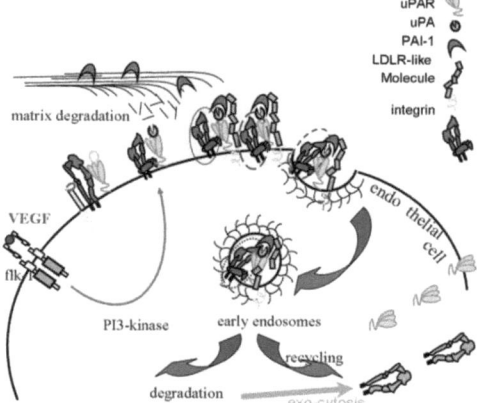

Figure 13. Schematic representation of uPAR mediated integrin internalization for VEGF stimulated endothelial cells.

The urokinase receptor can associate with integrin heterodimers in a cell-type dependent manner. The original observation was made with β2 integrins in complex with αL and αM; these associate with uPAR on monocytes (67). Formation of complexes between uPAR, integrins and an LDL-related receptor was first documented in HT1080 fibrosarcoma cells (128). Ligation of the LDL-related receptor protein-1 with PAI-1 drives the internalization of integrins αvβ3 and αvβ5 in conjunction with uPAR. Earlier experiments also suggested a functional interaction of α5β1 integrin and uPAR: upon increasing expression of uPAR, integrin α5β1 was recovered in the immunoprecipitate and this correlated with persistent ERK1/2 activation and enhanced tumor growth *in vivo* (97,98). We employed a micropatterning approach that allowed for direct visualization of VEGF-promoted formation of a complex between uPAR and α5β1 integrin on the endothelial cell surface. The salient techniques that are used to probe for protein interactions, such as co-immunoprecipitation or tandem affinity purification have several disadvantages. Protein interactions are probed in detergent lysed cell lysates where the spatio-temporal distribution of cell membrane proteins as well as their native confirmation is lost. Such settings are prone to false positives as well as false negatives where weak interactions can be lost. Endothelial cells seeded on the integrin antibody functionalized micropattern surfaces provide a platform for analyzing interactions occurring on the surface of live cells. The fact that integrin internalization is precluded in this experimental setup did not only facilitate quantitative assessment of uPAR recruitment but also afforded the unequivocal demonstration that the interaction happened indeed at the cell surface. This new approach does not allow to differentiate whether there is a direct interaction between uPAR and integrins or whether additional proteins - other than LDL receptor-like proteins - must be recruited to stabilize and internalize the complex. It is worth noting that purified uPAR and purified α5β1 integrin can directly interact in detergent solution (108).

Dynamin and clathrin-mediated endocytosis of various integrins has been documented in several instances (68-70). LDL-receptor like proteins internalize uPAR into clathrin coated vesicles (136). Experiments from our lab show that VEGF stimulation of endothelial cells stimulates internalization of β1 integrins into clathrin-coated vesicles. Therefore the hypothesis that uPAR and β1 integrins are internalizied by LDL-related proteins via clathrin coated vesicles is concurring with the observation that uPAR and β1 integrins were found in the same endocytic vesicles following VEGF stimulation. Growth factors involved in wound healing trigger the redistributive cycle of integrin trafficking to control cell migration (88,89). The downstream signalling pathways are not well understood. However, in endothelial cells, VEGF-A-triggered αvβ3 integrin recycling is contingent on activation of PKD1 (143). Our observations in signaling pathway studies are in line with our findings: integrin internalization was triggered by the receptor specific for VEGF-E

(FLK1/VEGFR-2) and blunted by wortmannin. This compound inhibits PI3-kinase, which is also the upstream activator of PKD1. Our earlier observation showed that VEGF induced a PI3-kinase dependent activation of MMP-2 (139). VEGF induced recycling of uPAR was contingent on this activation of MMP-2 and prevented by an inhibitor of MMP2/9 (140). Accordingly, we used this inhibitor interrogate the VEGF-controlled signaling network. These experiments further supported our conclusion that the VEGFR-2 orchestrates both, the accumulation of uPAR and a repertoire of proteolytic enzymes at the leading edge of an invading endothelial cell and regulated retrieval of β1 integrins via the same signaling pathway. Accordingly, uPAR is an essential bottleneck through which the VEGF-generated signal must be funnelled: in the absence of uPAR, neither focused proteolysis nor integrin-dependent migration can be elicited by VEGF.

Our observations indicate that the uPAR tri-molecular complex removes inactivated β1 integrins from the cell surface and in effect permitting migratory αvβ3 integrin to be exported to the cell surface. We have previously shown that VEGF stimulation leads to β1 integrin inactivation through a PI3-kinase pathway (140). Our hypothesis is supported by two facts. UPAR did not show any interaction with the activating β1 integrin antibody patterned microbiochip. pFAK associated with the β1 integrin reduced following VEGF stimulation. The absence of pFAK with uPAR and β1 integrin complexes implies that the integrin is in the inactive state.

In endothelial cells, VEGF triggered endocytosis of α3β1 integrin. Previous experiments documented that the binding of α3β1 integrin to uPAR occurred in a uPA-dependent manner (147). The direct binding of purified uPAR to purified α3β1 integrin and the recovery of uPAR in complex with α3β1 integrin is contingent on the presence of uPA (147,101). Surprisingly, micropatterning did not detect any direct interaction of α3β1 integrin with uPAR. In our opinion, micropatterning is among the most sensitive methods to record interactions in the native cell membrane, because it does not require any modification of the interacting molecules (e.g., by attaching fluorescent moieties) or any change in their stoichiometry (e.g., by heterologous expression). Thus, we conclude that, in endothelial cells, VEGF does not promote complex formation between α3β1 integrin und uPAR. This conclusion is also supported by the observation that VEGF-triggered endocytosis of α3β1 integrin was not impaired by RAP. The discrepancy between our findings and results from earlier studies (147,101) is most likely accounted for by cell type-dependent differences. In fact, uPAR can recruit various integrins (96-104) in a cell type-dependent manner. This also supports the conjecture that these complexes are stabilized by the recruitment of additional proteins.

In our model the VEGF-induced signal is propagated via uPAR to regulate the fibronectin receptor - α5β1 integrin. In the absence of uPAR (PI-PLC treatment or uPAR inhibitory peptides) VEGF induced internalization of α5β1 integrin is inhibited. Preventing the internalization of uPAR with RAP inhibits α5β1 integrin internalization convincing that it's internalization is contingent on uPAR. This subsequently also affects VEGF induced migration on collagen and fibronectin. The significance of our observations was proved with the inhibition of *in vivo* angiogenesis using the murine equivalent of the uPAR inhibitory peptide. In the presence of the inhibitory peptide VEGF induced endothelial invasion was significantly reduced with no visible blood vessel formation. This predicts and proves that the absence of α5β1 integrin ought to suppress tumor angiogenesis. Mice rendered genetically deficient in integrin α5 integrin die during embryonic development (86,87). As uPAR knock out animals are also viable the phenotype of mice with double knockout of α5 integrin and uPAR in endothelial cells would be interesting. However, blockage of α5β1 integrin by an antibody suppressed angiogenesis in a murine tumor model, where human rhabdomyosarcoma cells were xenografted into immunodeficient mice (148). Accordingly, α5β1 integrin is being explored as a potential target in advanced human cancer; the pertinent chimeric antibody, is currently in phase II clinical trials (149).

The binding sites for α5β1 integrin are thought to reside on domain-3 of uPAR. Specifically, point mutation of S245 or H249 resulted in inhibition of the association of uPAR and α5β1 integrin (99,100). This allows for selective disruption of the complex, a concept that was verified by using a peptide comprising the residues 243-251 of uPAR (TASMCQHAH) that contains both S245 and H249. The peptide efficiently blocked VEGF-induced internalization of the fibronectin receptor. The interaction site between uPAR and α5β1 integrin fulfils several criteria of a candidate drug binding site: (i) it is readily accessible, because it is on the extracellular surface thus obviating the cell membrane as a permeation barrier. (ii) It allows for discrimination, because it can be specifically targeted. (iii) There is no major toxicity that can be a priori anticipated. The absence of uPAR allows for the development of a viable animal (64) but it interferes with VEGF-induced angiogenesis. Our results therefore provide a proof-of-principle that the interface of uPAR and α5β1 integrin may represent a site to be targeted for anti-angiogenic therapy.

5. BIBLIOGRAPHY

1. Folkman J. Tumor angiogenesis: therapeutic implications. *N Engl J Med.* 1971;285(21):1182-6.
2. Carmeliet P. Angiogenesis in health and disease. *Nat Med.* 2003;9(6):653-60.
3. Vlodavsky I, Folkman J, Sullivan R, Fridman R, Ishai-Michaeli R, Sasse J, Klagsbrun M. Endothelial cell-derived basic fibroblast growth factor: synthesis and deposition into subendothelial extracellular matrix. *Proc Natl Acad Sci U S A.* 1987;84(8):2292-6.
4. Baird A, Esch F, Mormède P, Ueno N, Ling N, Böhlen P, Ying SY, Wehrenberg WB, Guillemin R. Molecular characterization of fibroblast growth factor: distribution and biological activities in various tissues. *Recent Prog Horm Res.* 1986;42:143-205.
5. Dennis PA, Rifkin DB. Studies on the role of basic fibroblast growth factor in vivo: inability of neutralizing antibodies to block tumor growth. *J Cell Physiol.* 1990;144(1):84-98.
6. Leung DW, Cachianes G, Kuang WJ, Goeddel DV, Ferrara N. Vascular endothelial growth factor is a secreted angiogenic mitogen. *Science.* 1989;246(4935):1306-9.
7. Keck PJ, Hauser SD, Krivi G, Sanzo K, Warren T, Feder J, Connolly DT. Vascular permeability factor, an endothelial cell mitogen related to PDGF. *Science.* 1989;246(4935):1309-12.
8. Millauer B, Wizigmann-Voos S, Schnürch H, Martinez R, Møller NP, Risau W, Ullrich A. High affinity VEGF binding and developmental expression suggest Flk-1 as a major regulator of vasculogenesis and angiogenesis. *Cell.* 1993;72(6):835-46.
9. Breier G, Albrecht U, Sterrer S, Risau W. Expression of vascular endothelial growth factor during embryonic angiogenesis and endothelial cell differentiation. *Development.* 1992;114(2):521-32.
10. Plate KH, Breier G, Weich HA, Risau W. Vascular endothelial growth factor is a potential tumour angiogenesis factor in human gliomas in vivo. *Nature.* 1992;359(6398):845-8.
11. Dvorak HF, Sioussat TM, Brown LF, Berse B, Nagy JA, Sotrel A, Manseau EJ, Van de Water L, Senger DR. Distribution of vascular permeability factor (vascular endothelial growth factor) in tumors: concentration in tumor blood vessels. *J Exp Med.* 1991;174(5):1275-8.
12. Shweiki D, Itin A, Soffer D, Keshet E. Vascular endothelial growth factor induced by hypoxia may mediate hypoxia-initiated angiogenesis. *Nature.* 1992;359(6398):843-5.

13. Liu Y, Cox SR, Morita T, Kourembanas S. Hypoxia regulates vascular endothelial growth factor gene expression in endothelial cells. Identification of a 5' enhancer. *Circ Res.* 1995;77(3):638-43.
14. Olsson AK, Dimberg A, Kreuger J, Claesson-Welsh L. VEGF receptor signalling - in control of vascular function. *Nat Rev Mol Cell Biol.* 2006;7(5):359-71.
15. Meyer M, Clauss M, Lepple-Wienhues A, Waltenberger J, Augustin HG, Ziche M, Lanz C, Büttner M, Rziha HJ, Dehio C. A novel vascular endothelial growth factor encoded by Orf virus, VEGF-E, mediates angiogenesis via signalling through VEGFR-2 (KDR) but not VEGFR-1 (Flt-1) receptor tyrosine kinases. *EMBO J.* 1999;18(2):363-74.
16. Kukk E, Lymboussaki A, Taira S, Kaipainen A, Jeltsch M, Joukov V, Alitalo K. VEGF-C receptor binding and pattern of expression with VEGFR-3 suggests a role in lymphatic vascular development. *Development.* 1996;122(12):3829-37.
17. Shalaby F, Rossant J, Yamaguchi TP, Gertsenstein M, Wu XF, Breitman ML, Schuh AC. Failure of blood-island formation and vasculogenesis in Flk-1-deficient mice. *Nature.* 1995;376(6535):62-6.
18. Fong GH, Rossant J, Gertsenstein M, Breitman ML. Role of the Flt-1 receptor tyrosine kinase in regulating the assembly of vascular endothelium. *Nature.* 1995;376(6535):66-70.
19. Waltenberger J, Claesson-Welsh L, Siegbahn A, Shibuya M, Heldin CH. Different signal transduction properties of KDR and Flt1, two receptors for vascular endothelial growth factor. *J Biol Chem.* 1994;269(43):26988-95.
20. Keyt BA, Nguyen HV, Berleau LT, Duarte CM, Park J, Chen H, Ferrara N.Identification of vascular endothelial growth factor determinants for binding KDR and FLT-1 receptors. Generation of receptor-selective VEGF variants by site-directed mutagenesis. *J Biol Chem.* 1996;271(10):5638-46.
21. Sakurai Y, Ohgimoto K, Kataoka Y, Yoshida N, Shibuya M. Essential role of Flk-1 (VEGF receptor 2) tyrosine residue 1173 in vasculogenesis in mice. *Proc Natl Acad Sci U S A.* 2005;102(4):1076-81.
22. Takahashi T, Yamaguchi S, Chida K, Shibuya M. A single autophosphorylation site on KDR/Flk-1 is essential for VEGF-A-dependent activation of PLC-gamma and DNA synthesis in vascular endothelial cells. *EMBO J.* 2001;20(11):2768-78.
23. Takahashi T, Ueno H, Shibuya M. VEGF activates protein kinase C-dependent, but Ras-independent Raf-MEK-MAP kinase pathway for DNA synthesis in primary endothelial cells. *Oncogene.* 1999;18(13):2221-30.

24. Holmqvist K, Cross MJ, Rolny C, Hägerkvist R, Rahimi N, Matsumoto T, Claesson-Welsh L, Welsh M. The adaptor protein shb binds to tyrosine 1175 in vascular endothelial growth factor (VEGF) receptor-2 and regulates VEGF-dependent cellular migration. *J Biol Chem.* 2004;279(21):22267-75.
25. Fujio Y, Walsh K. Akt mediates cytoprotection of endothelial cells by vascular endothelial growth factor in an anchorage-dependent manner. *J Biol Chem.* 1999;274(23):16349-54.
26. Gerber HP, McMurtrey A, Kowalski J, Yan M, Keyt BA, Dixit V, Ferrara N. Vascular endothelial growth factor regulates endothelial cell survival through the phosphatidylinositol 3'-kinase/Akt signal transduction pathway. Requirement for Flk-1/KDR activation. *J Biol Chem.* 1998;273(46):30336-43.
27. Collen D. Ham-Wasserman lecture: role of the plasminogen system in fibrin-homeostasis and tissue remodeling. *Hematology Am Soc Hematol Educ Program.* 2001:1-9.
28. Ellis V, Scully MF, Kakkar VV. Plasminogen activation initiated by single-chain urokinase-type plasminogen activator. Potentiation by U937 monocytes. *J Biol Chem.* 1989;264(4):2185-8.
29. Lijnen HR, Van Hoef B, Nelles L, Collen D. Plasminogen activation with single-chain urokinase-type plasminogen activator (scu-PA). Studies with active site mutagenized plasminogen (Ser740----Ala) and plasmin-resistant scu-PA (Lys158----Glu). *J Biol Chem.* 1990;265(9):5232-6.
30. Ellis V, Scully MF, Kakkar VV. Plasminogen activation by single-chain urokinase in functional isolation. A kinetic study. *J Biol Chem.* 1987 Nov 5;262(31):14998-5003.
31. Plow EF, Herren T, Redlitz A, Miles LA, Hoover-Plow JL. The cell biology of the plasminogen system. *FASEB J.* 1995;9(10):939-45.
32. Moses MA. The regulation of neovascularization by matrix metalloproteinases and their inhibitors. *Stem Cells.* 1997;15:180–189.
33. Libby P, Schönbeck U. Drilling for oxygen: angiogenesis involves proteolysis of the extracellular matrix. *Circ Res.* 2001;89(3):195-7.
34. Zucker S, Mirza H, Conner CE, Lorenz AF, Drews MH, Bahou WF, Jesty J. Vascular endothelial growth factor induces tissue factor and matrix metalloproteinase production in endothelial cells: conversion of prothrombin to thrombin results in progelatinase A activation and cell proliferation. *Int J Cancer.* 1998;75:780–786
35. Iwasaka C, Tanaka K, Abe M, Sato Y.Ets-1 regulates angiogenesis by inducing the expression of urokinase-type plasminogen activator and matrix metalloproteinase-1 and the migration of vascular endothelial cells. *J Cell Physiol.* 1996;169(3):522-31

36. Cubellis MV, Nolli ML, Cassani G, Blasi F. Binding of single-chain prourokinase to the urokinase receptor of human U937 cells. *J Biol Chem.* 1986;261(34):15819-22.
37. Ploug M, Rahbek-Nielsen H, Nielsen PF, Roepstorff P, Dano K. Glycosylation profile of a recombinant urokinase-type plasminogen activator receptor expressed in Chinese hamster ovary cells. *J Biol Chem.* 1998;273(22):13933-43.
38. Møller LB, Pöllänen J, Rønne E, Pedersen N, Blasi F. N-linked glycosylation of the ligand-binding domain of the human urokinase receptor contributes to the affinity for its ligand. *J Biol Chem.* 1993;268(15):11152-9.
39. Ploug M. Identification of specific sites involved in ligand binding by photoaffinity labeling of the receptor for the urokinase-type plasminogen activator. Residues located at equivalent positions in uPAR domains I and III participate in the assembly of a composite ligand-binding site. *Biochemistry.* 1998;37(47):16494-505.
40. Llinas P, Le Du MH, Gårdsvoll H, Danø K, Ploug M, Gilquin B, Stura EA, Ménez A. Crystal structure of the human urokinase plasminogen activator receptor bound to an antagonist peptide. *EMBO J.* 2005;24(9):1655-63.
41. Høyer-Hansen G, Rønne E, Solberg H, Behrendt N, Ploug M, Lund LR, Ellis V, Danø K. Urokinase plasminogen activator cleaves its cell surface receptor releasing the ligand-binding domain. *J Biol Chem.* 1992;267(25):18224-9.
42. Høyer-Hansen G, Ploug M, Behrendt N, Rønne E, Danø K. Cell-surface acceleration of urokinase-catalyzed receptor cleavage. *Eur J Biochem.* 1997;243(1-2):21-6.
43. Fazioli F, Resnati M, Sidenius N, Higashimoto Y, Appella E, Blasi F. A urokinase-sensitive region of the human urokinase receptor is responsible for its chemotactic activity. *EMBO J.* 1997;16(24):7279-86.
44. Stephens RW, Nielsen HJ, Christensen IJ, Thorlacius-Ussing O, Sørensen S, Danø K, Brünner N.Plasma urokinase receptor levels in patients with colorectal cancer: relationship to prognosis. *J Natl Cancer Inst.* 1999;91(10):869-74.
45. Pedersen H, Brünner N, Francis D, Osterlind K, Rønne E, Hansen HH, Danø K, Grøndahl-Hansen J. Prognostic impact of urokinase, urokinase receptor, and type 1 plasminogen activator inhibitor in squamous and large cell lung cancer tissue. *Cancer Res.* 1994;54(17):4671-5.
46. Mustjoki S, Sidenius N, Sier CF, Blasi F, Elonen E, Alitalo R, Vaheri A. Soluble urokinase receptor levels correlate with number of circulating tumor cells in acute myeloid leukemia and decrease rapidly during chemotherapy. *Cancer Res.* 2000;60(24):7126-32.

47. Riisbro R, Christensen IJ, Piironen T, Greenall M, Larsen B, Stephens RW, Han C, Høyer-Hansen G, Smith K, Brünner N, Harris AL.Prognostic significance of soluble urokinase plasminogen activator receptor in serum and cytosol of tumor tissue from patients with primary breast cancer. *Clin Cancer Res.* 2002;8(5):1132-41.
48. Barnathan ES, Kuo A, Rosenfeld L, Karikó K, Leski M, Robbiati F, Nolli ML, Henkin J, Cines DB. Interaction of single-chain urokinase-type plasminogen activator with human endothelial cells. *J Biol Chem.* 1990;265(5):2865-72.
49. Reidy MA, Irvin C, Lindner V. Migration of arterial wall cells. Expression of plasminogen activators and inhibitors in injured rat arteries. *Circ Res.* 1996;78(3):405-14.
50. Yamamoto M, Sawaya R, Mohanam S, Rao VH, Bruner JM, Nicolson GL, Rao JS. Expression and localization of urokinase-type plasminogen activator receptor in human gliomas. *Cancer Res.* 1994;54(18):5016-20.
51. Mandriota SJ, Seghezzi G, Vassalli JD, Ferrara N, Wasi S, Mazzieri R, Mignatti P, Pepper MS.Vascular endothelial growth factor increases urokinase receptor expression in vascular endothelial cells. *J Biol Chem.* 1995;270(17):9709-16.
52. Pepper MS, Sappino AP, Stöcklin R, Montesano R, Orci L, Vassalli JD. Upregulation of urokinase receptor expression on migrating endothelial cells. *J Cell Biol.* 1993;122(3):673-84.
53. Smith HW, Marshall CJ. Regulation of cell signalling by uPAR. *Nat Rev Mol Cell Biol.* 2010;11(1):23-36.
54. Min HY, Doyle LV, Vitt CR, Zandonella L, Stratton-Thomas JR, Shuman MA, Rosenberg S. Urokinase receptor antagonists inhibit angiogenesis and primary tumor growth in syngeneic mice. *Cancer Res.* 1996;56:2428–2433
55. Evans CP, Elfman F, Parangi S, Conn M, Cunha G, Shuman MA. Inhibition of prostate cancer neovascularization and growth by urokinase-plasminogen activator receptor blockade. *Cancer Res.* 1997;57:3594–3599
56. Crowley CW, Cohen RL, Lucas BK, Liu G, Shuman MA, Levinson AD.Prevention of metastasis by inhibition of the urokinase receptor. *Proc Natl Acad Sci U S A.* 1993;90(11):5021-5.
57. Dewerchin M, Nuffelen AV, Wallays G, Bouché A, Moons L, Carmeliet P, Mulligan RC, Collen D.Generation and characterization of urokinase receptor-deficient mice. *J Clin Invest.* 1996;97(3):870-8.
58. Carmeliet P, Moons L, Dewerchin M, Rosenberg S, Herbert JM, Lupu F, Collen D. Receptor-independent role of urokinase-type plasminogen activator in pericellular plasmin

and matrix metalloproteinase proteolysis during vascular wound healing in mice. *J Cell Biol.* 1998;140(1):233-45.

59. Levi M, Moons L, Bouché A, Shapiro SD, Collen D, Carmeliet P. Deficiency of urokinase-type plasminogen activator-mediated plasmin generation impairs vascular remodeling during hypoxia-induced pulmonary hypertension in mice. *Circulation.* 2001;103(15):2014-20.

60. Bugge TH, Flick MJ, Danton MJ, Daugherty CC, Romer J, Dano K, Carmeliet P, Collen D, Degen JL.Urokinase-type plasminogen activator is effective in fibrin clearance in the absence of its receptor or tissue-type plasminogen activator. *Proc Natl Acad Sci U S A.* 1996;93(12):5899-904.

61. Bajou K, Masson V, Gerard RD, Schmitt PM, Albert V, Praus M, Lund LR, Frandsen TL, Brunner N, Dano K, Fusenig NE, Weidle U, Carmeliet G, Loskutoff D, Collen D, Carmeliet P, Foidart JM, Noël A. The plasminogen activator inhibitor PAI-1 controls in vivo tumor vascularization by interaction with proteases, not vitronectin. Implications for antiangiogenic strategies. *J Cell Biol. 2001*;152(4):777-84.

62. Carmeliet P, Moons L, Herbert JM, Crawley J, Lupu F, Lijnen R, Collen D. Urokinase but not tissue plasminogen activator mediates arterial neointima formation in mice. *Circ Res.* 1997;81(5):829-39.

63. Carmeliet P, Schoonjans L, Kieckens L, Ream B, Degen J, Bronson R, De Vos R, van den Oord JJ, Collen D, Mulligan RC. Physiological consequences of loss of plasminogen activator gene function in mice. *Nature.* 1994;368(6470):419-24.

64. May AE, Kanse SM, Lund LR, Gisler RH, Imhof BA, Preissner KT. Urokinase receptor (CD87) regulates leukocyte recruitment via beta 2 integrins in vivo. *J Exp Med.* 1998;188(6):1029-37.

65. Gyetko MR, Sud S, Kendall T, Fuller JA, Newstead MW, Standiford TJ. Urokinase receptor-deficient mice have impaired neutrophil recruitment in response to pulmonary Pseudomonas aeruginosa infection. *J Immunol.* 2000;165(3):1513-9.

66. Rijneveld AW, Levi M, Florquin S, Speelman P, Carmeliet P, van Der Poll T.Urokinase receptor is necessary for adequate host defense against pneumococcal pneumonia. *J Immunol.* 2002;168(7):3507-11.

67. Bohuslav J, Horejsí V, Hansmann C, Stöckl J, Weidle UH, Majdic O, Bartke I, Knapp W, Stockinger H. Urokinase plasminogen activator receptor, beta 2-integrins, and Src-kinases within a single receptor complex of human monocytes. *J Exp Med.* 1995;181(4):1381-90.

68. Pellinen T, Ivaska J. Integrin traffic. *J Cell Sci.* 2006;119(Pt 18):3723-31.

69. Caswell PT, Vadrevu S, Norman JC. Integrins: masters and slaves of endocytic transport. *Nat Rev Mol Cell Biol.* 2009;10(12):843-53.
70. Pellinen T, Tuomi S, Arjonen A, Wolf M, Edgren H, Meyer H, Grosse R, Kitzing T, Rantala JK, Kallioniemi O, Fässler R, Kallio M, Ivaska J. Integrin trafficking regulated by Rab21 is necessary for cytokinesis. *Dev Cell.* 2008;15(3):371-85.
71. Valdembri D, Caswell PT, Anderson KI, Schwarz JP, König I, Astanina E, Caccavari F, Norman JC, Humphries MJ, Bussolino F, Serini G. Neuropilin-1/GIPC1 signaling regulates alpha5beta1 integrin traffic and function in endothelial cells. *PLoS Biol.* 2009;7(1):e25.
72. Chao WT, Kunz J. Focal adhesion disassembly requires clathrin-dependent endocytosis of integrins. *FEBS Lett.* 2009;583(8):1337-43.
73. Sincock PM, Fitter S, Parton RG, Berndt MC, Gamble JR, Ashman LK. PETA-3/CD151, a member of the transmembrane 4 superfamily, is localised to the plasma membrane and endocytic system of endothelial cells, associates with multiple integrins and modulates cell function. *J Cell Sci.* 1999;112 (Pt 6):833-44.
74. Davis GE. Affinity of integrins for damaged extracellular matrix: alpha v beta 3 binds to denatured collagen type I through RGD sites. *Biochem Biophys Res Commun.* 1992;182(3):1025-31.
75. Xu J, Rodriguez D, Petitclerc E, Kim JJ, Hangai M, Moon YS, Davis GE, Brooks PC. Proteolytic exposure of a cryptic site within collagen type IV is required for angiogenesis and tumor growth in vivo. *J Cell Biol.* 2001;154(5):1069-79.
76. Rüegg C, Mariotti A. Vascular integrins: pleiotropic adhesion and signaling molecules in vascular homeostasis and angiogenesis. *Cell Mol Life Sci.* 2003;60(6):1135-57.
77. Brooks PC, Clark RA, Cheresh DA. Requirement of vascular integrin alpha v beta 3 for angiogenesis. *Science.* 1994;264(5158):569-71.
78. Stupack DG, Cheresh DA. ECM remodeling regulates angiogenesis: endothelial integrins look for new ligands. *Sci STKE.* 2002;2002(119):pe7.
79. Avraamides CJ, Garmy-Susini B, Varner JA. Integrins in angiogenesis and lymphangiogenesis. *Nat Rev Cancer.* 2008;8(8):604-17.
80. Friedlander M, Brooks PC, Shaffer RW, Kincaid CM, Varner JA, Cheresh DA. Definition of two angiogenic pathways by distinct alpha v integrins. *Science.* 1995;270(5241):1500-2.
81. Hodivala-Dilke KM, McHugh KP, Tsakiris DA, Rayburn H, Crowley D, Ullman-Culleré M, Ross FP, Coller BS, Teitelbaum S, Hynes RO. Beta3-integrin-deficient mice are a model for Glanzmann thrombasthenia showing placental defects and reduced survival. *J Clin Invest.* 1999;103(2):229-38.

82. Huang X, Griffiths M, Wu J, Farese RV Jr, Sheppard D. Normal development, wound healing, and adenovirus susceptibility in beta5-deficient mice. *Mol Cell Biol.* 2000;20(3):755-9.
83. Reynolds LE, Wyder L, Lively JC, Taverna D, Robinson SD, Huang X, Sheppard D, Hynes RO, Hodivala-Dilke KM.Enhanced pathological angiogenesis in mice lacking beta3 integrin or beta3 and beta5 integrins. *Nat Med.* 2002;8(1):27-34.
84. Bader BL, Rayburn H, Crowley D, Hynes RO.Extensive vasculogenesis, angiogenesis, and organogenesis precede lethality in mice lacking all alpha v integrins. *Cell.* 1998;95(4):507-19.
85. Tanjore H, Zeisberg EM, Gerami-Naini B, Kalluri R. Beta1 integrin expression on endothelial cells is required for angiogenesis but not for vasculogenesis. *Dev Dyn.* 2008;237(1):75-82.
86. Yang JT, Rayburn H, Hynes RO.Embryonic mesodermal defects in alpha 5 integrin-deficient mice. *Development.* 1993;119(4):1093-105.
87. van der Flier A, Badu-Nkansah K, Whittaker CA, Crowley D, Bronson RT, Lacy-Hulbert A, Hynes RO. Endothelial alpha5 and alphav integrins cooperate in remodeling of the vasculature during development. *Development.* 2010;137(14):2439-49.
88. Roberts M, Barry S, Woods A, van der Sluijs P, Norman J. PDGF-regulated rab4-dependent recycling of alphavbeta3 integrin from early endosomes is necessary for cell adhesion and spreading. *Curr Biol.* 2001;11(18):1392-402.
89. Thomas M, Felcht M, Kruse K, Kretschmer S, Deppermann C, Biesdorf A, Rohr K, Benest AV, Fiedler U, Augustin HG. Angiopoietin-2 stimulation of endothelial cells induces alphavbeta3 integrin internalization and degradation. *J Biol Chem.* 2010;285(31):23842-9.
90. Reynolds AR, Hart IR, Watson AR, Welti JC, Silva RG, Robinson SD, Da Violante G, Gourlaouen M, Salih M, Jones MC, Jones DT, Saunders G, Kostourou V, Perron-Sierra F, Norman JC, Tucker GC, Hodivala-Dilke KM. Stimulation of tumor growth and angiogenesis by low concentrations of RGD-mimetic integrin inhibitors. *Nat Med.* 2009;15(4):392-400.
91. Waltz DA, Sailor LZ, Chapman HA. Cytokines induce urokinase-dependent adhesion of human myeloid cells. A regulatory role for plasminogen activator inhibitors. *J Clin Invest.* 1993;91(4):1541-52.
92. Wei Y, Waltz DA, Rao N, Drummond RJ, Rosenberg S, Chapman HA. Identification of the urokinase receptor as an adhesion receptor for vitronectin. *J Biol Chem.* 1994;269(51):32380-8.

93. Gårdsvoll H, Ploug M. Mapping of the vitronectin-binding site on the urokinase receptor: involvement of a coherent receptor interface consisting of residues from both domain I and the flanking interdomain linker region. *J Biol Chem.* 2007;282(18):13561-72.
94. Madsen CD, Ferraris GM, Andolfo A, Cunningham O, Sidenius N. uPAR-induced cell adhesion and migration: vitronectin provides the key. *J Cell Biol.* 2007;177(5):927-39.
95. Salasznyk RM, Zappala M, Zheng M, Yu L, Wilkins-Port C, McKeown-Longo PJ. The uPA receptor and the somatomedin B region of vitronectin direct the localization of uPA to focal adhesions in microvessel endothelial cells. *Matrix Biol.* 2007;26(5):359-70.
96. Wei Y, Lukashev M, Simon DI, Bodary SC, Rosenberg S, Doyle MV, Chapman HA. Regulation of integrin function by the urokinase receptor. *Science.* 1996;273(5281):1551-5.
97. Aguirre Ghiso JA, Kovalski K, Ossowski L. Tumor dormancy induced by downregulation of urokinase receptor in human carcinoma involves integrin and MAPK signaling. *J Cell Biol.* 1999;147(1):89-104.
98. Aguirre Ghiso JA. Inhibition of FAK signaling activated by urokinase receptor induces dormancy in human carcinoma cells in vivo. *Oncogene.* 2002;21(16):2513-24.
99. Xue W, Mizukami I, Todd RF 3rd, Petty HR. Urokinase-type plasminogen activator receptors associate with beta1 and beta3 integrins of fibrosarcoma cells: dependence on extracellular matrix components. *Cancer Res.* 1997;57(9):1682-9.
100. Tarui T, Mazar AP, Cines DB, Takada Y. Urokinase-type plasminogen activator receptor (CD87) is a ligand for integrins and mediates cell-cell interaction. *J Biol Chem.* 2001;276(6):3983-90
101. Wei Y, Eble JA, Wang Z, Kreidberg JA, Chapman HA. Urokinase receptors promote beta1 integrin function through interactions with integrin alpha3beta1. *Mol Biol Cell.* 2001;12(10):2975-86.
102. Wei C, Möller CC, Altintas MM, Li J, Schwarz K, Zacchigna S, Xie L, Henger A, Schmid H, Rastaldi MP, Cowan P, Kretzler M, Parrilla R, Bendayan M, Gupta V, Nikolic B, Kalluri R, Carmeliet P, Mundel P, Reiser J. Modification of kidney barrier function by the urokinase receptor. *Nat Med.* 2008;14(1):55-63.
103. Smith HW, Marra P, Marshall CJ. uPAR promotes formation of the p130Cas-Crk complex to activate Rac through DOCK180. *J Cell Biol.* 2008;182(4):777-90.
104. Kjøller L, Hall A. Rac mediates cytoskeletal rearrangements and increased cell motility induced by urokinase-type plasminogen activator receptor binding to vitronectin. *J Cell Biol.* 2001;152(6):1145-57.

105. Simon DI, Wei Y, Zhang L, Rao NK, Xu H, Chen Z, Liu Q, Rosenberg S, Chapman HA. Identification of a urokinase receptor-integrin interaction site. Promiscuous regulator of integrin function. *J Biol Chem.* 2000;275(14):10228-34.
106. Zhang F, Tom CC, Kugler MC, Ching TT, Kreidberg JA, Wei Y, Chapman HA. Distinct ligand binding sites in integrin alpha3beta1 regulate matrix adhesion and cell-cell contact. *J Cell Biol.* 2003;163(1):177-88.
107. Ghosh S, Johnson JJ, Sen R, Mukhopadhyay S, Liu Y, Zhang F, Wei Y, Chapman HA, Stack MS. Functional relevance of urinary-type plasminogen activator receptor-alpha3beta1 integrin association in proteinase regulatory pathways. *J Biol Chem.* 2006;281(19):13021-9.
108. Chaurasia P, Aguirre-Ghiso JA, Liang OD, Gardsvoll H, Ploug M, Ossowski L. A region in urokinase plasminogen receptor domain III controlling a functional association with alpha5beta1 integrin and tumor growth. *J Biol Chem.* 2006;281(21):14852-63.
109. Wei Y, Tang CH, Kim Y, Robillard L, Zhang F, Kugler MC, Chapman HA. Urokinase receptors are required for alpha 5 beta 1 integrin-mediated signaling in tumor cells. *J Biol Chem.* 2007;282(6):3929-39.
110. Tang CH, Hill ML, Brumwell AN, Chapman HA, Wei Y. Signaling through urokinase and urokinase receptor in lung cancer cells requires interactions with beta1 integrins. *J Cell Sci.* 2008;121(Pt 22):3747-56.
111. Schiller HB, Szekeres A, Binder BR, Stockinger H, Leksa V. Mannose 6-phosphate/insulin-like growth factor 2 receptor limits cell invasion by controlling alphaVbeta3 integrin expression and proteolytic processing of urokinase-type plasminogen activator receptor. *Mol Biol Cell.* 2009;20(3):745-56.
112. Cheng X, Shen Z, Yin L, Lu SH, Cui Y. ECRG2 regulates cell migration/invasion through urokinase-type plasmin activator receptor (uPAR)/beta1 integrin pathway. *J Biol Chem.* 2009;284(45):30897-906.
113. Liu D, Aguirre Ghiso J, Estrada Y, Ossowski L. EGFR is a transducer of the urokinase receptor initiated signal that is required for in vivo growth of a human carcinoma. *Cancer Cell.* 2002;1(5):445-57.
114. Herz J, Couthier DE, Hammer RE. Correction: LDL receptor-related protein internalizes and degrades uPA-PAI-1 complexes and is essential for embryo implantation. *Cell.* 1993;73(3):428.
115. Cao C, Lawrence DA, Li Y, Von Arnim CA, Herz J, Su EJ, Makarova A, Hyman BT, Strickland DK, Zhang L. Endocytic receptor LRP together with tPA and PAI-1 coordinates Mac-1-dependent macrophage migration. *EMBO J.* 2006;25(9):1860-70.

116. Bu G, Maksymovitch EA, Schwartz AL. Receptor-mediated endocytosis of tissue-type plasminogen activator by low density lipoprotein receptor-related protein on human hepatoma HepG2 cells. *J Biol Chem.* 1993;268(17):13002-9.

117. Holst-Hansen C, Johannessen B, Høyer-Hansen G, Rømer J, Ellis V, Brünner N. Urokinase-type plasminogen activation in three human breast cancer cell lines correlates with their in vitro invasiveness. *Clin Exp Metastasis.* 1996;14(3):297-307.

118. Webb DJ, Thomas KS, Gonias SL. Plasminogen activator inhibitor 1 functions as a urokinase response modifier at the level of cell signaling and thereby promotes MCF-7 cell growth. *J Cell Biol.* 2001;152(4):741-52.

119. Geetha N, Mihaly J, Stockenhuber A, Blasi F, Uhrin P, Binder BR, Freissmuth M, Breuss JM. Signal integration and coincidence detection in the mitogen-activated protein kinase/ERK-cascade: Concomitant activation of receptor tyrosine kinases and of LRP-1 leads to sustained ERK phosphorylation via downregulation of dual specificity phosphatases (DUSP-1 and -6). *J Biol Chem.* 2011;286(29):25663-74.

120. Muratoglu SC, Mikhailenko I, Newton C, Migliorini M, Strickland DK. Low density lipoprotein receptor-related protein 1 (LRP1) forms a signaling complex with platelet-derived growth factor receptor-beta in endosomes and regulates activation of the MAPK pathway. *J Biol Chem.* 2010;285(19):14308-17.

121. Newton CS, Loukinova E, Mikhailenko I, Ranganathan S, Gao Y, Haudenschild C, Strickland DK. Platelet-derived growth factor receptor-beta (PDGFR-beta) activation promotes its association with the low density lipoprotein receptor-related protein (LRP). Evidence for co-receptor function. *J Biol Chem.* 2005;280(30):27872-8.

122. Qi JH, Claesson-Welsh L. VEGF-induced activation of phosphoinositide 3-kinase is dependent on focal adhesion kinase. *Exp Cell Res.* 2001;263(1):173-82.

123. Loskutoff DJ, Curriden SA, Hu G, Deng G. Regulation of cell adhesion by PAI-1. *APMIS.* 1999;107(1):54-61.

124. Seiffert D, Loskutoff DJ. Evidence that type 1 plasminogen activator inhibitor binds to the somatomedin B domain of vitronectin. *J Biol Chem.* 1991;266(5):2824-30.

125. Okumura Y, Kamikubo Y, Curriden SA, Wang J, Kiwada T, Futaki S, Kitagawa K, Loskutoff DJ. Kinetic analysis of the interaction between vitronectin and the urokinase receptor. *J Biol Chem.* 2002;277(11):9395-404.

126. Deng G, Curriden SA, Wang S, Rosenberg S, Loskutoff DJ. Is plasminogen activator inhibitor-1 the molecular switch that governs urokinase receptor-mediated cell adhesion and release? *J Cell Biol.* 1996;134(6):1563-71.

127. Deng G, Curriden SA, Hu G, Czekay RP, Loskutoff DJ. Plasminogen activator inhibitor-1 regulates cell adhesion by binding to the somatomedin B domain of vitronectin. *J Cell Physiol.* 2001;189(1):23-33.

128. Czekay RP, Aertgeerts K, Curriden SA, Loskutoff DJ. Plasminogen activator inhibitor-1 detaches cells from extracellular matrices by inactivating integrins. *J Cell Biol.* 2003;160(5):781-91.

129. Czekay RP, Loskutoff DJ. Plasminogen activator inhibitors regulate cell adhesion through a uPAR-dependent mechanism. *J Cell Physiol.* 2009;220(3):655-63.

130. Wei Y, Czekay RP, Robillard L, Kugler MC, Zhang F, Kim KK, Xiong JP, Humphries MJ, Chapman HA. Regulation of alpha5beta1 integrin conformation and function by urokinase receptor binding. *J Cell Biol.* 2005;168(3):501-11.

131. Olson D, Pöllänen J, Høyer-Hansen G, Rønne E, Sakaguchi K, Wun TC, Appella E, Danø K, Blasi F. Internalization of the urokinase-plasminogen activator inhibitor type-1 complex is mediated by the urokinase receptor. *J Biol Chem.* 1992;267(13):9129-33.

132. Cubellis MV, Wun TC, Blasi F. Receptor-mediated internalization and degradation of urokinase is caused by its specific inhibitor PAI-1. *EMBO J.* 1990;9(4):1079-85.

133. Nykjaer A, Kjøller L, Cohen RL, Lawrence DA, Garni-Wagner BA, Todd RF 3rd, van Zonneveld AJ, Gliemann J, Andreasen PA. Regions involved in binding of urokinase-type-1 inhibitor complex and pro-urokinase to the endocytic alpha 2-macroglobulin receptor/low density lipoprotein receptor-related protein. Evidence that the urokinase receptor protects pro-urokinase against binding to the endocytic receptor. *J Biol Chem.* 1994;269(41):25668-76.

134. Conese M, Nykjaer A, Petersen CM, Cremona O, Pardi R, Andreasen PA, Gliemann J, Christensen EI, Blasi F. alpha-2 Macroglobulin receptor/Ldl receptor-related protein(Lrp)-dependent internalization of the urokinase receptor. *J Cell Biol.* 1995;131(6 Pt 1):1609-22.

135. Stefansson S, Muhammad S, Cheng XF, Battey FD, Strickland DK, Lawrence DA. Plasminogen activator inhibitor-1 contains a cryptic high affinity binding site for the low density lipoprotein receptor-related protein. *J Biol Chem.* 1998;273(11):6358-66.

136. Czekay RP, Kuemmel TA, Orlando RA, Farquhar MG. Direct binding of occupied urokinase receptor (uPAR) to LDL receptor-related protein is required for endocytosis of uPAR and regulation of cell surface urokinase activity. *Mol Biol Cell.* 2001;12(5):1467-79.

137. Nykjaer A, Conese M, Christensen EI, Olson D, Cremona O, Gliemann J, Blasi F. Recycling of the urokinase receptor upon internalization of the uPA:serpin complexes. *EMBO J.* 1997;16(10):2610-20.

138. Prager GW, Breuss JM, Steurer S, Mihaly J, Binder BR. Vascular endothelial growth factor (VEGF) induces rapid prourokinase (pro-uPA) activation on the surface of endothelial cells. *Blood.* 2004;103(3):955-62.

139. Argraves KM, Battey FD, MacCalman CD, McCrae KR, Gåfvels M, Kozarsky KF, Chappell DA, Strauss JF 3rd, Strickland DK. The very low density lipoprotein receptor mediates the cellular catabolism of lipoprotein lipase and urokinase-plasminogen activator inhibitor type I complexes. *J Biol Chem.* 1995;270(44):26550-7.

140. Prager GW, Breuss JM, Steurer S, Olcaydu D, Mihaly J, Brunner PM, Stockinger H, Binder BR. Vascular endothelial growth factor receptor-2-induced initial endothelial cell migration depends on the presence of the urokinase receptor. *Circ Res.* 2004;94(12):1562-70.

141. Hohenegger M, Mitterauer M, Voss T, Nanoff C, Freissmuth M. Thiophosphorylation of the G protein beta subunit in human platelet membranes : Evidence against a direct phosphate transfer reaction to G alpha subunits. *Mol Pharmacol.* 1996;49(1):73-80.

142. Schwarzenbacher M, Kaltenbrunner M, Brameshuber M, Hesch C, Paster W, Weghuber J, Heise B, Sonnleitner A, Stockinger H, Schütz GJ. Micropatterning for quantitative analysis of protein-protein interactions in living cells. *Nat Methods.* 2008;5(12):1053-60.

143. Blasio L di, Droetto S, Norman J, Bussolino F, Primo L. Protein kinase D1 regulates VEGF-a-induced alphavbeta3 integrin trafficking and endothelial cell migration. *Traffic* 2010;11(8):1107-18.

144. Balsara RD, Merryman R, Virjee F, Northway C, Castellino FJ, Ploplis VA. A deficiency of uPAR alters endothelial angiogenic function and cell morphology. *Vasc Cell.* 2011;3(1):10.

145. Pulukuri SM, Gondi CS, Lakka SS, Jutla A, Estes N, Gujrati M, Rao JS. RNA interference-directed knockdown of urokinase plasminogen activator and urokinase plasminogen activator receptor inhibits prostate cancer cell invasion, survival, and tumorigenicity in vivo. *J Biol Chem.* 2005;280(43):36529-40.

146. Gondi CS, Lakka SS, Dinh DH, Olivero WC, Gujrati M, Rao JS. RNAi-mediated inhibition of cathepsin B and uPAR leads to decreased cell invasion, angiogenesis and tumor growth in gliomas. *Oncogene.* 2004;23(52):8486-96.

147. Mazzieri R, D'Alessio S, Kenmoe RK, Ossowski L, Blasi F. An uncleavable uPAR mutant allows dissection of signaling pathways in uPA-dependent cell migration. *Mol Biol Cell* 2006;17:367-378.

148. Bhaskar V, Zhang D, Fox M, Seto P, Wong MH, Wales PE, Powers D, Chao DT, Dubridge RB, Ramakrishnan V. A function blocking anti-mouse integrin alpha5beta1 antibody inhibits angiogenesis and impedes tumor growth in vivo. *J Transl Med.* 2007 Nov;5:61

149. Bell-McGuinn KM, Matthews CM, Ho SN, Barve M, Gilbert L, Penson RT, Lengyel E, Palaparthy R, Gilder K, Vassos A, McAuliffe W, Weymer S, Barton J, Schilder RJ. A phase II, single-arm study of the anti-α5β1 integrin antibody volociximab as monotherapy in patients with platinum-resistant advanced epithelial ovarian or primary peritoneal cancer. *Gynecol Oncol.* 2011;121(2):273-9.

150. Hiratsuka S, Minowa O, Kuno J, Noda T, Shibuya M. Flt-1 lacking the tyrosine kinase domain is sufficient for normal development and angiogenesis in mice. *Proc Natl Acad Sci U S A.* 1998; 95(16):9349-54.

151. Murakami M, Iwai S, Hiratsuka S, Yamauchi M, Nakamura K, Iwakura Y, Shibuya M. Signaling of vascular endothelial growth factor receptor-1 tyrosine kinase promotes rheumatoid arthritis through activation of monocytes/macrophages. *Blood.* 2006;108(6):1849-56.

152. Resnati M, Pallavicini I, Wang JM, Oppenheim J, Serhan CN, Romano M, Blasi F. The fibrinolytic receptor for urokinase activates the G protein-coupled chemotactic receptor FPRL1/LXA4R. *Proc Natl Acad Sci U S A.* 2002;99(3):1359-64.

153. Montuori N, Carriero MV, Salzano S, Rossi G, Ragno P. The cleavage of the urokinase receptor regulates its multiple functions. *J Biol Chem.* 2002 ;277(49):46932-9.

154. Gao W, Wang Z, Bai X, Xi X, Ruan C. Detection of soluble urokinase receptor by immunoradiometric assay and its application in tumor patients. *Thromb Res.* 2001;102(1):25-31.

i want morebooks!

Buy your books fast and straightforward online - at one of world's fastest growing online book stores! Environmentally sound due to Print-on-Demand technologies.

Buy your books online at
www.get-morebooks.com

Kaufen Sie Ihre Bücher schnell und unkompliziert online – auf einer der am schnellsten wachsenden Buchhandelsplattformen weltweit! Dank Print-On-Demand umwelt- und ressourcenschonend produziert.

Bücher schneller online kaufen
www.morebooks.de

VDM Verlagsservicegesellschaft mbH
Heinrich-Böcking-Str. 6-8
D - 66121 Saarbrücken

Telefon: +49 681 3720 174
Telefax: +49 681 3720 1749

info@vdm-vsg.de
www.vdm-vsg.de

Printed by Books on Demand GmbH, Norderstedt / Germany